Ion Traps

A gentle introduction

Online at: https://doi.org/10.1088/978-0-7503-5472-1

Ion Traps

A gentle introduction

Masatoshi Kajita
Terahertz Technology Research Center, National Institute of Information and Communications Technology, Tokyo, Japan

IOP Publishing, Bristol, UK

ISBN 978-0-7503-5472-1 (ebook)
ISBN 978-0-7503-5470-7 (print)
ISBN 978-0-7503-5473-8 (myPrint)
ISBN 978-0-7503-5471-4 (mobi)

DOI 10.1088/978-0-7503-5472-1

Version: 20221201

IOP ebooks

British Library Cataloguing-in-Publication Data: A catalogue record for this book is available from the British Library.

Published by IOP Publishing, wholly owned by The Institute of Physics, London

IOP Publishing, No.2 The Distillery, Glassfields, Avon Street, Bristol, BS2 0GR, UK

US Office: IOP Publishing, Inc., 190 North Independence Mall West, Suite 601, Philadelphia, PA 19106, USA

Contents

Preface

Detailed research into particles (atoms, molecules, etc) is possible by observing the same particles continuously for a long period. For this purpose, we need some force for trapping without changing the properties of the trapped particles. The motion of ions can be manipulated by the electric or magnetic field. Three-dimensional trapping of an ion is possible using a combination of the DC magnetic field and the inhomogeneous DC electric field (Penning trap). A Penning trap is useful for the precision measurement of mass ratios between different charged particles.

Using an inhomogeneous RF electric field, ions are tightly trapped in a small area for a long period of time (RF trap, Paul trap). The motion energy of the trapped ion is reduced to the lowest eigen states by laser cooling. The internal quantum energy state of the trapped ion can be also manipulated. Therefore, we can observe the phenomena of one or a few ions with a deterministic quantum state (eigen state or coupled state). For example, a single ion was observed at two positions simultaneously (Schroedinger's cat). The precision measurement of transition frequencies of RF trapped ions has been performed and the fractional uncertainty of 10^{-18} was obtained with a few transitions.

This book introduces the fundamentals of ion trap technology and their use for the development of new physics, chemistry, or engineering and is targetted at graduate course students.

The development of the ion trap was the subject of the Nobel prizes:

1989 H G Dehmelt and W Paul

2012 D J Wineland

Acknowledgement

The research activity of the author is supported by a Grant-in-Aid for Scientific Research (B) (Grant Nos. JP17H02881 and JP20H01920), and a Grant-in-Aid for Scientific Research (C) (Grant Nos. JP17K06483 and 16K05500) from the Japan Society for the Promotion of Science (JSPS). The author is highly appreciative of discussions with Y Yano, T Ido, N Ohtsubo, A Shinjo-Kihara, H Hachisu, S Nagano, M Kumagai, S Hayashi, N Sekine, and M Hara, all from the NICT, Japan, as well as K Okada (Sophia University), T Aoki (the University of Tokyo), and N Kimura (RIKEN). The author is grateful to K Kameta and J Navas (IOP, UK) for the opportunity to write this book.

Author biography

Masatoshi Kajita

Born and raised in Nagoya, Japan, Dr Kajita graduated from the Department of Applied Physics, the University of Tokyo in 1981 and obtained his PhD from the Department of Physics, the University of Tokyo in 1986. After working at the Institute for Molecular Science, he joined the Communications Research Laboratory (CRL) in 1989. In 2004, the CRL was renamed the National Institute of Information and Communications Technology (NICT). In 2009, he was a guest professor at the Université de Provence, Marseille, France.

IOP Publishing

Ion Traps
A gentle introduction
Masatoshi Kajita

Chapter 1

What is an ion trap?

This chapter introduces the principle of an ion trap. We can manipulate the motion of ions using the electromagnetic field. We can trap ions in a small spatial area using the combination of a DC magnetic field and a DC electric field (Penning trap) or inhomogeneous AC electric field (RF trap).

Different configurations of electrodes were developed to give proper distributions of electric field.

1.1 Introduction

In the gaseous phase, atoms and molecules exist in isolated states. Note that the properties of single atoms and molecules are uniform everywhere, while the properties of solids or fluids cannot be uniform because the bonding between the composing atoms and molecules depends on the circumstance and some impurities are always included. Therefore, research into gaseous atoms and molecules is useful to understand the fundamentals of physics. For example, precision measurement of atomic transition frequencies made it possible to develop atomic clocks with measurement uncertainties that are much lower than those for crystal clocks [1].

However, gaseous particles (atoms and molecules) move at high speed (several hundred m s^{-1}). We cannot observe the same particles continuously because the particles in an observation area (e.g., the area under the irradiation of probe laser light) change rapidly as shown in figure 1.1(a). When we observe the gaseous particles in a vessel, they collide with the wall and change the state. We can observe the average phenomena of particles with different states (velocity and internal quantum energy state discussed in chapter 2). To examine the properties of particles in a selected state in detail, it is preferable to trap them in a small spatial region without changing their quantum state. Development of laser cooling (section 2.4) made it possible to reduce the kinetic energy of particles to below 1 mK [2]. Ultra-cold particles can be trapped in a small area using a standing wave of laser light or inhomogeneous magnetic field [3, 4]. However, the depth of the trap potential is

doi:10.1088/978-0-7503-5472-1ch1

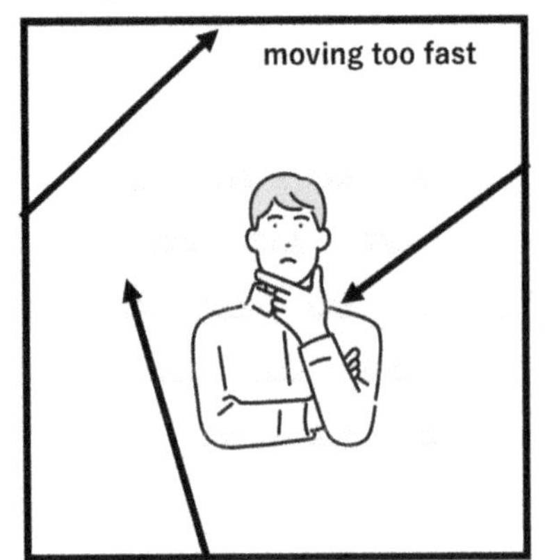

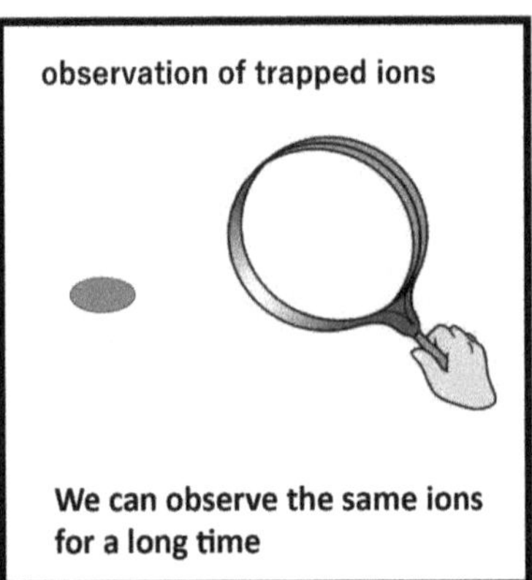

Figure 1.1. (a) Gaseous particles (atoms or molecules) at room temperature move very fast and change the energy state by colliding. We can observe the phenomena as the average of many particles with different energy states. We cannot observe the phenomena of particles in a selected quantum state. (b) For ions trapped in a small spatial region, we can observe the phenomena of the same ion in a selected quantum state taking a long time.

below 1 mK and the period of trapping is limited to be shorter than 1 s by the collision with background gas (kinetic energy of 300 K).

On the other hand, charged particles have strong interactions with the electromagnetic field and their motion can be manipulated much more easily than neutral particles. For example, charged particles have been accelerated using the electromagnetic field [5]. An ion trap is a technology used to trap ions in a small spatial region by interaction with the electromagnetic field, as shown in figure 1.1(b). The trapping period is much longer than those for neutral atoms or molecules because the trapping potential is much deeper than the kinetic energy of background gas. We can even observe a single trapped ion continuously. Observing the phenomena of trapped ions for a long time, we can investigate quantum phenomena which cannot be described by classical mechanics. We can also measure the transition frequencies of trapped ions with ultra-low uncertainty.

1.2 Equation of motion of ions and Maxwell's equation

A moving mass with a velocity $\vec{v}$ and an electric charge of q_e experiences a force due to an electric field $\vec{E}$ and a magnetic field $\vec{B}$ as follows [6].

$$
\begin{aligned}
\vec{F} &= q_e[\vec{E} + \vec{v} \times \vec{B}] \\
F_x &= q_e E_x + v_y B_z - v_z B_y \\
F_y &= q_e E_y + v_z B_x - v_x B_z \\
F_z &= q_e E_z + v_x B_y - v_y B_x.
\end{aligned}
\tag{1.2.1}
$$

The distribution of the electromagnetic field is derived using Maxwell's equations shown below (in SI units) [6]

$\nabla \cdot \vec{E} = \frac{\rho}{\varepsilon}$ (r: the electric charge density, ε: the permittivity)

$$\frac{\partial E_x}{\partial x} + \frac{\partial E_y}{\partial y} + \frac{\partial E_z}{\partial z} = \frac{\rho}{\varepsilon} \tag{1.2.2}$$

$$\nabla \times \vec{E} = -\frac{\partial \vec{B}}{\partial t}$$

$$\frac{\partial E_z}{\partial y} - \frac{\partial E_y}{\partial z} = -\frac{\partial B_x}{\partial t}, \quad \frac{\partial E_x}{\partial z} - \frac{\partial E_z}{\partial x} = -\frac{\partial B_y}{\partial t}, \quad \frac{\partial E_y}{\partial x} - \frac{\partial E_x}{\partial y} = -\frac{\partial B_z}{\partial t} \tag{1.2.3}$$

$$\nabla \cdot \vec{B} = 0$$

$$\frac{\partial B_x}{\partial x} + \frac{\partial B_y}{\partial y} + \frac{\partial B_z}{\partial z} = 0 \tag{1.2.4}$$

$\nabla \times \vec{B} = \mu\left[\vec{j} + \varepsilon\frac{\partial \vec{B}}{\partial t}\right]$ ($\vec{j}$: the electric current density, μ: the permeability)

$$\frac{\partial B_z}{\partial y} - \frac{\partial B_y}{\partial z} = \mu\left[j_x + \varepsilon\frac{\partial E_x}{\partial t}\right], \quad \frac{\partial B_x}{\partial z} - \frac{\partial B_z}{\partial x} = \mu\left[j_y + \varepsilon\frac{\partial E_y}{\partial t}\right],$$
$$\frac{\partial B_y}{\partial x} - \frac{\partial B_x}{\partial y} = \mu\left[j_z + \varepsilon\frac{\partial E_z}{\partial t}\right]. \tag{1.2.5}$$

Considering the vacuum space, equation (1.2.2) is rewritten as

$$\frac{\partial E_x}{\partial x} + \frac{\partial E_y}{\partial y} + \frac{\partial E_z}{\partial z} = 0 \tag{1.2.6}$$

1.3 Can we trap ions using only a DC electric field?

First, we consider the possibility of trapping a charged particle at a point $(x, y, z) = (0,0,0)$ using an inhomogeneous DC electric field.

To trap in the x direction, $q_eE_x > 0$ at $x < 0$ and $q_eE_x < 0$ at $x > 0$ is required, therefore, $q_e\frac{dE_x}{dx} < 0$ is required, as shown in figure 1.2(a). To trap in all directions, $q_e\frac{dE_y}{dy} < 0$ and $q_e\frac{dE_z}{dz} < 0$ are also required. To trap ions in all directions,

$$q_e\left(\frac{\partial E_x}{\partial x} + \frac{\partial E_y}{\partial y} + \frac{\partial E_z}{\partial z}\right) < 0. \tag{1.3.2}$$

is required, and this is not possible from equation (1.2.6). Using just the inhomogeneous DC electric field, ions can be trapped in one or two directions but cannot be trapped in three directions.

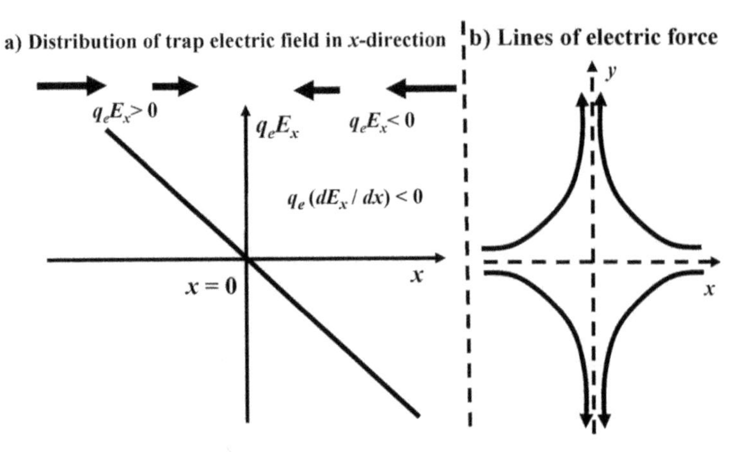

Figure 1.2. (a) Indicates that $q_e(dE_x/dx) < 0$ is required for the trapping in the x direction and (b) indicates that the DC electric field, giving a trap force in one direction, must have an expansion force in another direction, because the line of electric force cannot have squirt or suction.

Figure 1.2(b) indicates a line of electric force, which cannot have a squirt or suction in the vacuum. When the electric field gives a trapping force in the x direction, the line of electric force at $x < 0$ and $x > 0$ must have inverse directions. The line of electric force is bent at around $x = 0$ and an expansion force is given in a different direction.

1.4 Trap using DC magnetic field and DC electric field (Penning trap)

When a DC magnetic field B_z is applied in the z direction, the equation of motion of an ion with the electric charge of q_c and mass of m in the xy-plane is given by (see equation (1.2.1))

$$m\frac{d^2x}{dt^2} = q_eB_z\frac{dy}{dt} \tag{1.4.1}$$

$$m\frac{d^2y}{dt^2} = -q_eB_z\frac{dx}{dt}. \tag{1.4.2}$$

Adding both sides of equation (1.4.1) and $\lambda \times$ equation (1.4.2)

$$m\frac{d^2(x + \lambda y)}{dt^2} = q_eB_z\frac{d(-\lambda x + y)}{dt}$$

when $1: \lambda = -\lambda: 1$ $\lambda = \pm i$

$$m\frac{d^2(x \pm iy)}{dt^2} = \mp \; iq_eB_z\frac{d(x \pm iy)}{dt}$$

$$\frac{d(x \pm iy)}{dt} = c_{\pm}\exp\left(\mp i\frac{q_eB_z}{m}t\right)\left(\frac{dx}{dt}\right)^2 + \left(\frac{dr}{dt}\right)^2 = c_+c_- = v_0^2 \text{ (constant)}$$

$$\frac{dx}{dt} = -v_0\sin\left(\frac{q_eB_z}{m}t + \vartheta_0\right)\frac{dy}{dt} = v_0\cos\left(\frac{q_eB_z}{m}t + \vartheta_0\right)$$

$$x = r_0\cos\left(\frac{q_eB_z}{m}t + \vartheta_0\right) + x_0 y = r_0\sin\left(\frac{q_eB_z}{m}t + \vartheta_0\right) + y_0 r_0 = \frac{m}{q_eB_z}v_0. \tag{1.4.3}$$

The ion has a circular motion with the frequency of $q_eB_z/2\pi m$ in the xy-plane as shown in figure 1.3. In other words, the ion is trapped in the region $x_0 - r_0 < x < x_0 + r_0$, $y_0 - r_0 < y < y_0 + r_0$. Equation (1.4.3) shows that the trap region is small for ions with small mass and high magnetic field.

Applying a DC electric field in the z direction with $q_eE_z > 0$ at $z < 0$ and $q_eE_z < 0$ at $z > 0$, ions are trapped in the vicinity of $z = 0$. When $E_z = \alpha_z z$, the motion in the z direction is given by

$$z = z_0\sin\left[\sqrt{\frac{q_e\alpha_z}{m}}t + \theta_z\right] \tag{1.4.4}$$

where z_0 and θ_z are the values given by the initial position and velocity.

Ions are trapped in three directions with the combination of a DC magnetic field and a DC electric field, which is called a Penning trap [7]. The Penning trap requires an ultra-high vacuum to suppress collisions with the background gas, because there is no force to fix the position of the center of the circle motion (x_0, y_0).

A Penning trap is applicable not only for ions, but also for all charged particles like electrons or protons. Equation (1.4.3) indicates that the frequency of the circle motion is inversely proportional to the mass of trapped charged particles (including

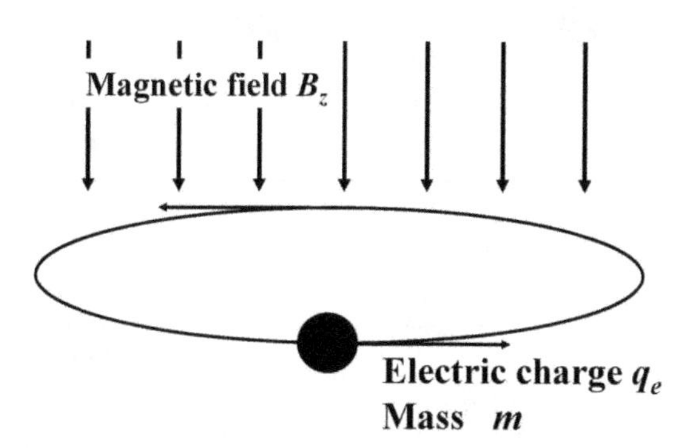

Figure 1.3. Circular motion of a charged particle in the xy-plane under a magnetic field in the z direction, this is called cyclotron motion.

ions), a Penning trap is often used as a precise mass spectrometer, as shown in section 5.9.

1.5 Fundamentals of an RF trap

Ions can be trapped using an inhomogeneous AC electric field. For an inhomogeneous RF electric field $\vec{E}(\vec{r})\sin(\Omega t)$, trapping and expanding forces are periodically applied. The time-averaged force can be a trapping force in all directions under certain conditions (RF trap). The equation of motion of a charged particle is given by

$$m\frac{d^2\vec{r}}{dt^2} = q_e\vec{E}(\vec{r})\sin(\Omega t). \tag{1.5.1}$$

When the change in position within the period of the RF electric field $\delta\vec{r}$ is negligibly small compared to $|\vec{r}|$, the time-averaged force is obtained to be non-zero as follows (see figure 1.4)

$$\vec{r} = \vec{r}_0 + \delta\vec{r}$$

$$m\frac{d^2\delta\vec{r}}{dt^2} = q_e\vec{E}(\vec{r}_0)\sin(\Omega t)$$

$$\delta\vec{r} = -\frac{q_e\vec{E}(\vec{r}_0)}{m\Omega^2}\sin(\Omega t) + c_1 t + c_0$$

($c_{0,1}$: constant given by the initial position and velocity)

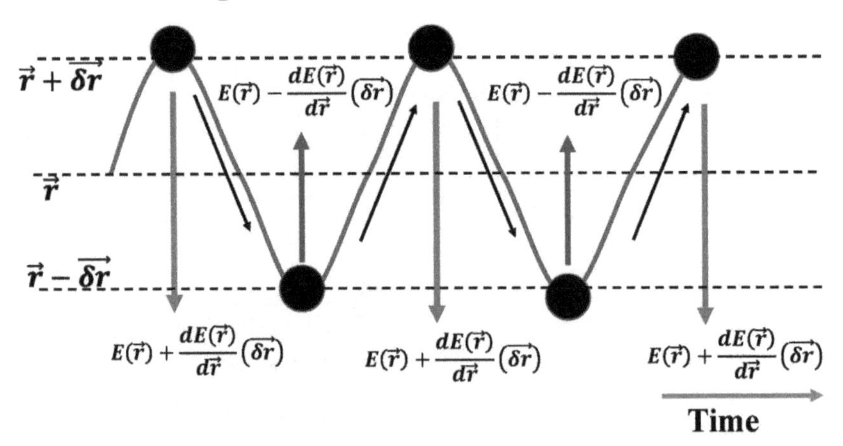

Figure 1.4. The mechanism to trap a charged particle using an AC electric field. The micromotion of the charged particle is synchronized to the AC electric field and a non-zero averaged force is induced in one direction.

$$\left[m\frac{d^2\vec{r}_0}{dt^2}\right]_{\text{ave}} = \left[q_e\vec{E}(\vec{r}_0 + \delta\vec{r})\sin(\Omega t)\right]_{\text{ave}}$$

$$= \left[q_e\vec{E}(\vec{r}_0)\sin(\Omega t)\right]_{\text{ave}} + \left[q_e\frac{d\vec{E}(\vec{r}_0)}{d\vec{r}}\delta\vec{r}\sin(\Omega t)\right]_{\text{ave}}$$

$$= -\left[q_e\frac{d\vec{E}(\vec{r}_0)}{d\vec{r}}\left\{\frac{q_e\vec{E}(\vec{r}_0)}{m\Omega^2}\sin(\Omega t)^2 + (c_1 t + c_0)\sin(\Omega t)\right\}\right]_{\text{ave}} \tag{1.5.2}$$

$$= -\frac{q_e^2\vec{E}(\vec{r}_0)}{2m\Omega^2}\frac{d\vec{E}(\vec{r}_0)}{d\vec{r}}.$$

The motion of ion (change of r_0) as represented by equation (1.5.2) is given as the motion with a pseudopotential field given by

$$P_{\text{ps}}(\vec{r}_0) = \frac{|\,q_e\vec{E}(\vec{r}_0)\,|^2}{4m\Omega^2}. \tag{1.5.3}$$

Equation (1.5.3) shows $P_{\text{ps}}(\vec{r}_0) \geqslant 0$. There is a force to confine the ion at the position where $|\,\vec{E}(\vec{r}_0)\,|$ is minimum and the trapping is also more stable than a Penning trap when the collision with background gas is significant. As shown in section 1.6, the minimum value of $|\,\vec{E}(\vec{r}_0)\,|$ is zero with most trapping electrodes.

Note that

$$|\,\vec{E}(\vec{r}_0)\,| \gg \left|\frac{d\vec{E}(\vec{r}_0)}{d\vec{r}}\delta\vec{r}\right| \to \frac{q_e}{m\Omega^2}\left|\frac{d\vec{E}(\vec{r}_0)}{d\vec{r}}\right| \ll 1 \tag{1.5.4}$$

is required when using equation (1.5.2). When equation (1.5.4) is not satisfied, the RF electric field can give an expanding force.

When equation (1.5.4) is satisfied, the motion of the trapped ion is separated to the secular motion induced by the pseudopotential (change of r_o) and the micromotion (change of δr). When $\vec{E} = (E_x, E_y, E_z)$ and $E_\rho = \alpha_\rho\rho(\rho = x, y, z)$ with constant values of α_ρ, the motion in the ρ-direction is given by (see figure 1.5)

$$\rho(t) = \mathrm{P}_0\sin[2\pi\nu_m t + \theta_s]\left(1 - \frac{q_e\alpha_\rho}{m\Omega^2}\sin(\Omega t)\right) \quad \mathrm{P}_0\text{: arbitrary constant}$$

$$\nu_m = \frac{q_e\alpha_\rho}{2\pi\sqrt{2}m\Omega}. \tag{1.5.5}$$

Equation (1.5.4) is satisfied with any value of P_0 when $q_e\alpha_\rho \ll m\Omega^2$. The mean kinetic energy of the secular motion is given by

$$\langle K\rangle_{\text{pseudo}} = \frac{m}{2}(P_0\omega)^2 = \frac{(q_e\alpha_\rho P_0)^2}{8m\Omega^2} \tag{1.5.6}$$

and that of the micromotion energy (kinetic energy) is given by

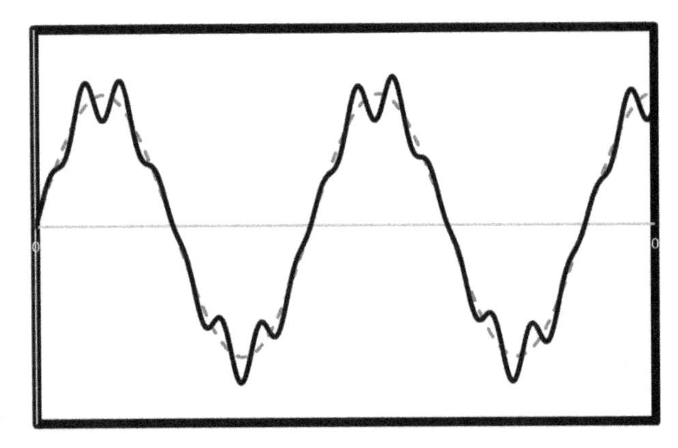

Figure 1.5. The motion of ion in an RF trap, given by equation (1.5.5).

$$\langle K\rangle_{\text{micro}} = \frac{m}{2}\left(\frac{q_e\alpha_\rho P_0}{m\Omega^2}\right)^2\Omega^2 = \frac{\left(q_e\alpha_\rho P_0\right)^2}{4m\Omega^2} = 2\langle K\rangle_{\text{pseudo}}. \tag{1.5.7}$$

The micromotion energy is larger than the secular motion energy, and it transforms to that of the secular motion by the collision between trapped ions. As the energy of the micromotion is supplied by the RF electric field after the transform to the secular motion energy, the total energy of the trapped ion is increased (heating). The energy conservation is not valid when there is a temporal change in the trap field. The heating effect also exists without the collision between trapped ions. When the relation between E_ρ and ρ is not linear the validity of equation (1.5.4) depends on the amplitude of the secular motion P_0. When equation (1.5.4) is not valid with large P_0, the motion is not given by equation (1.5.5) and the kinetic energy increases. The heating effect is often suppressed using buffer gas. The kinetic energy of ions is maintained at the same temperature as the buffer gas. The lower kinetic energy (smaller secular motion amplitude) is attained by laser cooling (see section 2.4).

Equations (1.5.5)–(1.5.7) show that the amplitude of the micromotion is reduced when the kinetic energy of the secular motion energy is reduced. With the real electrode, there is an extra electric charge at some position of the electrode and the real trap potential is not always minimum at the position with the zero RF electric field. Then the amplitude of the micromotion (kinetic energy of micromotion) is large also when $\langle K\rangle_{\text{pseudo}}$ is reduced. This problem is solved by adjusting the trap potential by adding a DC electric field (see figure 1.6), so that it becomes minimum at the position where the RF electric field is zero.

1.6 Different configurations of electrodes to give inhomogeneous electric field

The distribution of the electric field is convenient to consider the voltage Φ (when $\frac{\partial B}{\partial t} = 0$) as follows

$$\Phi = -\int E_x dx - \int E_y dy - \int E_z dz$$

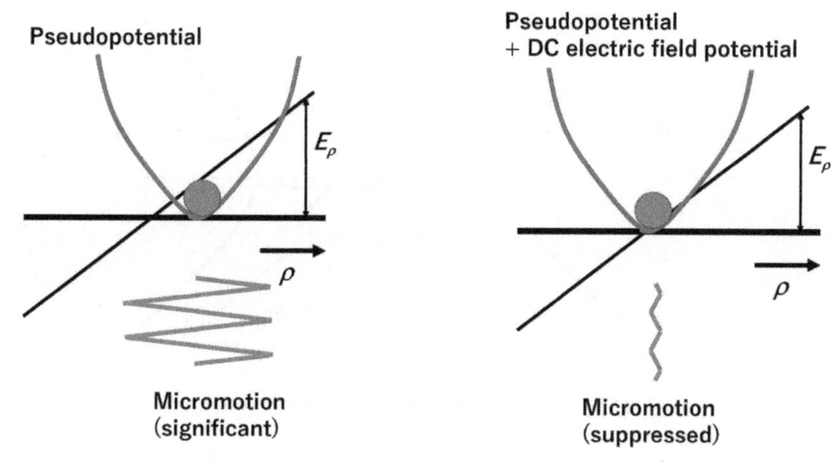

Figure 1.6. The pseudopotential is not always minimum with the position of zero AC trap electric field E_p. The micromotion is significant in this situation. Adjusting the position of the trap potential by an additional DC electric field, the micromotion is suppressed.

$$\frac{\partial^2\Phi}{\partial x^2} + \frac{\partial^2\Phi}{\partial y^2} + \frac{\partial^2\Phi}{\partial z^2} = 0, \tag{1.6.1}$$

because the boundary condition is given by the voltage on the electrode surface. For a Penning trap, the distribution of the electric field is not serious while $q_e E_z > 0$ at $z < 0$ and $q_e E_z < 0$ at $z > 0$ is satisfied. Therefore, we discuss this just for the application for the RF trap.

1.6.1 Electrodes for three-dimensional trapping

First, we consider the hyperboloid electrode with the surface condition (see figure 1.7)

$$\Phi = \Phi_0 \quad x^2 + y^2 - 2z^2 = -2z_0^2$$

$$\Phi = 0 \quad x^2 + y^2 - 2z^2 = r_0^2. \tag{1.6.2}$$

The voltage distribution is given by

$$\Phi(x, y, z) = -\frac{\Phi_0[x^2 + y^2 - 2z^2 - r_0^2]}{r_0^2 + 2z_0^2}. \tag{1.6.3}$$

The electric field is given by

$$E_x = \frac{2\Phi_0 x}{r_0^2 + 2z_0^2}, \quad E_y = \frac{2\Phi_0 y}{r_0^2 + 2z_0^2}, \quad E_z = -\frac{4\Phi_0 z}{r_0^2 + 2z_0^2}. \tag{1.6.4}$$

Here we consider the case that the AC and DC voltages are simultaneously applied as follows

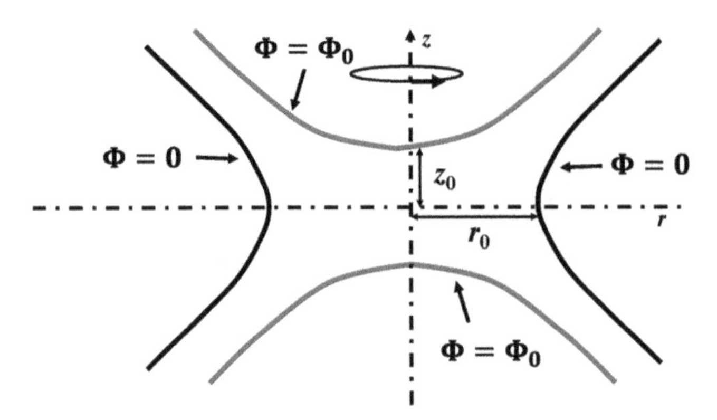

Figure 1.7. Configuration of hyperboloid electrode.

$$\Phi_0 = \Phi_{dc} + \Phi_{ac}\sin(\Omega t). \tag{1.6.5}$$

The amplitude of the micromotion is given by

$$\begin{pmatrix} \delta x \\ \delta y \\ \delta z \end{pmatrix} = \frac{2q_e}{m\Omega^2}\frac{\Phi_{ac}}{r_0^2 + 2z_0^2}\begin{pmatrix} x \\ y \\ -2z \end{pmatrix}, \tag{1.6.6}$$

and equation (1.5.4) is satisfied when

$$\frac{2q_e}{m\Omega^2}\frac{\Phi_{ac}}{r_0^2 + 2z_0^2} \ll 1. \tag{1.6.7}$$

The condition for the stable trapping shown by equation (1.5.4) is discussed independent of the position when the hyperboloid electrode is used. The pseudopotential induced by the RF electric field is given by

$$P_{\text{ps}}(\vec{r}) = \frac{1}{4m\Omega^2}\left(\frac{2q_e\Phi_{ac}}{r_0^2 + 2z_0^2}\right)^2(x^2 + y^2 + 4z^2), \tag{1.6.8}$$

and the total potential is given by

$$P_{\text{tot}}(\vec{r}) = -\frac{q_e\Phi_{dc}[x^2 + y^2 - 2z^2 - r_0^2]}{r_0^2 + 2z_0^2} + \frac{1}{4m\Omega^2}\left(\frac{2q_e\Phi_{ac}}{r_0^2 + 2z_0^2}\right)^2(x^2 + y^2 + 4z^2). \tag{1.6.9}$$

For the three-dimensional RF trapping, $P_{tot}(\vec{r}) \geqslant 0$ is required for all positions. The stability of the three-dimensional RF trap is discussed in detail using Mathieu's equation [8, 9]

$$\frac{d^2r}{d\tau^2} + \left[a_r^M + 2q_r^M\sin(2\tau)\right]r = 0,$$

$$r = \sqrt{x^2 + y^2}$$

$$\frac{d^2z}{d\tau^2} + [a_z^M + 2q_z^M \sin(2\tau)]z = 0$$

$$a_r^M = \frac{4q_e}{m\Omega^2}\frac{\Phi_{dc}}{r_0^2 2z_0^2}, \quad a_z^M = -2a_r^M \tag{1.6.10}$$

$$q_r^M = \frac{2q_e}{m\Omega^2}\frac{\Phi_{ac}}{r_0^2 + 2z_0^2}, \quad q_z^M = -2q_r^M, \quad \tau = \frac{\Omega t}{2}$$

Equation (1.6.7) is rewritten as

$$q_r^M \ll 1 \tag{1.6.11}$$

and the secular motion of the trapped ion is given by the harmonic oscillation with the frequencies of

$$x, y - \text{directions:} \quad \nu_m^{x,y} = \frac{\Omega}{2\pi}\sqrt{a_r^M + \frac{\left(q_r^M\right)^2}{2}}$$

$$z - \text{directions:} \quad \nu_m^z = \frac{\Omega}{2\pi}\sqrt{a_z^M + \frac{\left(q_z^M\right)^2}{2}}. \tag{1.6.12}$$

For the stable trapping, $a_{r,z}^M + \frac{\left(q_{r,z}^M\right)^2}{2} > 0$ is required to give the real values of the oscillation frequencies. The DC electric field is used to manipulate $(\nu_m^{x,y}/\nu_m^z)$, as a_r^M and a_z^M cannot be simultaneously positive. Without the DC electric field, stable trapping is obtained with $\left| q_{r,z}^M \right| < 0.908$. A practical experiment is performed mainly with $\left| q_{r,z}^M \right| < 0.3$, so that the amplitude of the micromotion is small enough to reduce the heating effect.

Although the hyperboloid electrode is convenient to discuss the motion stability of trapped ions using Mathieu's equation, it is not always realistic to prepare a hyperboloid electrode. For example, small electrodes are required to trap a single ion in a small spatial area. It is also advantageous to observe the fluorescence from the trapped ion. Therefore, trap electrodes with different configurations have been developed. For all electrodes, the electric field is zero at the trap center $(x, y, z) = (0,0,0)$ and the electric field distribution is approximately given by $\vec{E}(\vec{r}) = (\alpha_r x, \alpha_r y, \alpha_z z)$ at the vicinity of the trap center.

Figure 1.8 shows a ring electrode, which is often used to trap a single ion, whose configuration is much simpler than the hyperboloid electrode. Figure 1.9 shows the configuration of the endcap electrode, which is advantageous to observe the fluorescence from the trapped ion [10, 11].

1.6.2 Two-dimensional RF trapping + DC electric trapping

We can also trap ions with a combination of the two-dimensional (x,y directions) RF trapping and trapping in the z direction using a DC electric field. The AC electric field is uniform in the z direction and the AC voltage is given by

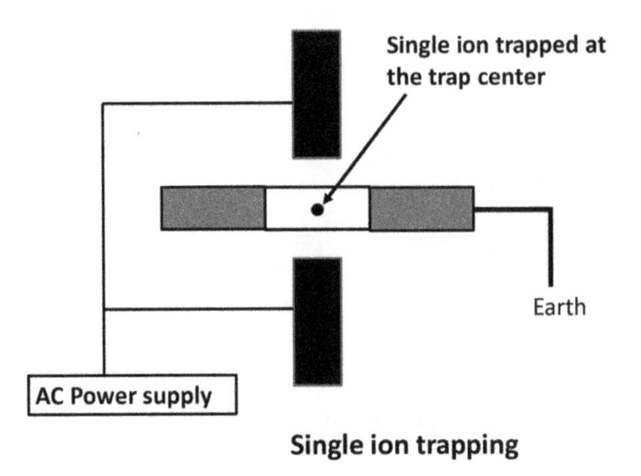

Figure 1.8. Configuration of a ring electrode, which is used to trap a single ion.

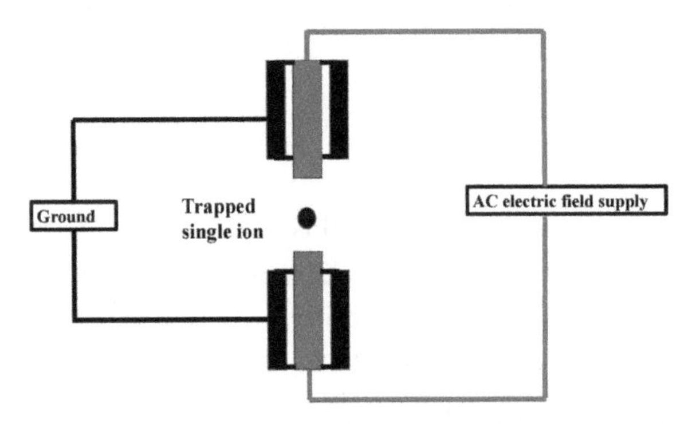

Figure 1.9. Confguration of an endcap electrode, which is used to trap a single ion.

$$\frac{\partial^2 \Phi}{\partial x^2} + \frac{\partial^2 \Phi}{\partial y^2} = 0. \tag{1.6.13}$$

Using a linear electrode with the hyperboloid cross-section

$$\Phi = \Phi_{ac}\sin(\Omega t) \qquad \text{at } x^2 - y^2 = r_0^2$$

$$\Phi = 0 \qquad \text{at } x^2 - y^2 = -r_0^2, \tag{1.6.14}$$

the voltage distribution is given by

$$\Phi(x, y) = \frac{[x^2 - y^2 + r_0^2]}{2r_0^2}\Phi_{ac}\sin(\Omega t) \tag{1.6.15}$$

and the electric field is

$$E_x^{ac} = \frac{x}{r_0^2}\Phi_{ac}\sin(\Omega t), \;\; E_y^{ac} = -\frac{y}{r_0^2}\Phi_{ac}\sin(\Omega t). \tag{1.6.16}$$

The AC trap electric field is zero on the z-axis ($(x, y) = (0,0)$).
The DC electric field satisfying

$$q_e E_z^{dc} > 0 \; z < 0, \;\; q_e E_z^{dc} < 0 \; z > 0 \left(q_e \frac{dE_z^{dc}}{dz} < 0 \right), \tag{1.6.17}$$

is applied using a pair of electrodes at $z = \pm z_0$. For technical reasons, the trapping is often performed using three parts of linear electrodes with the voltages:

$$-z_0 \leqslant z \leqslant z_0$$

$$\Phi = \Phi_{ac}\sin(\Omega t) \qquad \text{at } x^2 - y^2 = r_0^2$$

$$\Phi = 0 \qquad \text{at } x^2 - y^2 = -r_0^2$$

$$z_0 < | z | \leqslant z_0 + z_d$$

$$\Phi = \Phi_{ac}\sin(\Omega t) + \Phi_{dc} \qquad \text{at } x^2 - y^2 = r_0^2$$

$$\Phi = \Phi_{dc} \qquad \text{at } x^2 - y^2 = -r_0^2. \tag{1.6.18}$$

Note that the DC electric field giving a trap force in the z direction must give an expansion force to the x or y direction as shown in section 1.3. The RF trap force in the x, y directions must be larger than the expansion force by the DC electric field, as indicated by

$$\left| \frac{q_e^2 E_{x,\,y}^{ac}}{2m\Omega^2} \frac{dE_{x,\,y}^{ac}}{dx,\,y} \right| > \left| q_e E_{x,\,y}^{dc} \right|. \tag{1.6.19}$$

There is no micromotion in the z direction, therefore, the heating effect is much less significant than the three-dimensional RF trapping.

Here we consider the linear electrode with a hyperboloid cross-section for simplicity. The actual electrodes can be constructed using four cylindrical rods. The linear electrode with an endcap type (figure 1.9) cross-section is also used [12]. The planar linear electrode was also demonstrated, as shown in figure 1.10. The planar trap electrode is useful to make a chip scale ion trap [13].

Above, we introduced the quadrupole electrode. Aimed at the trapping of many ions, multipole linear electrodes using N_p ($\geqslant 4$) cylindrical rods were also developed [14]. Figure 1.11 shows a cross-section of the octupole linear trap electrode. We consider the xy-plane with the two-dimensional polar coordinate (r, θ). The cylindrical rods are given at $\theta = \frac{2n\pi}{N_p}$. An AC voltage is applied to the electrodes with even n and the ground is given to the electrodes with odd n. Equation (1.6.13) is rewritten as

$$\frac{\partial^2 \Phi}{\partial r^2} + \frac{1}{r}\frac{\partial \Phi}{\partial r} + \frac{1}{r^2}\frac{\partial^2 \Phi}{\partial \theta^2} = 0. \tag{1.6.20}$$

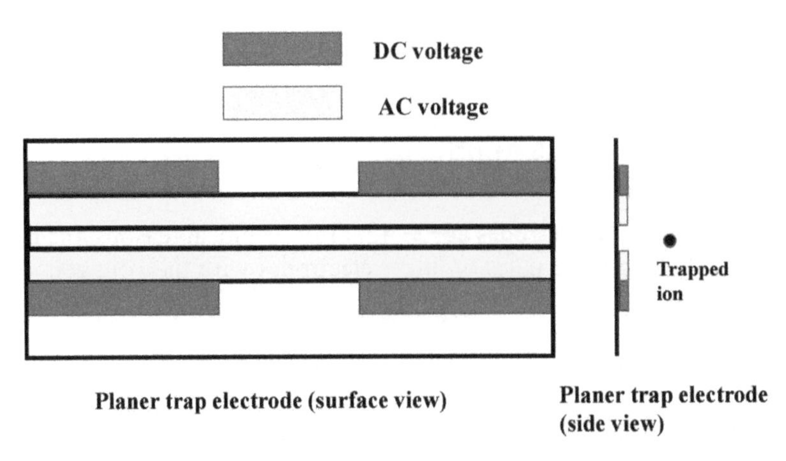

Figure 1.10. The configuration of a planar ion trap apparatus.

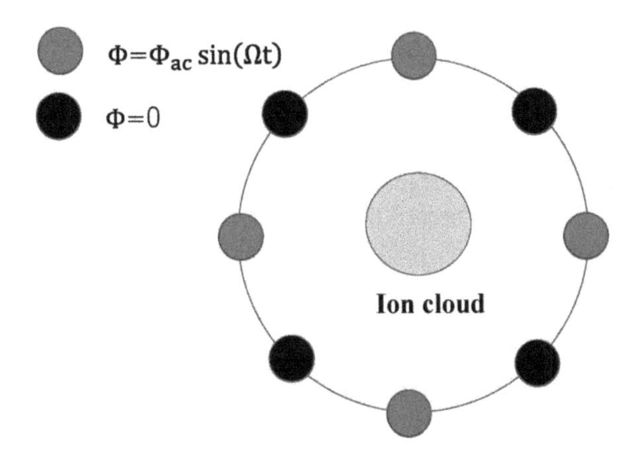

Figure 1.11. Cross-section of the octupole linear trap electrode.

From the periodicity of the voltage with θ,

$$\Phi \propto \cos\left(\frac{N_p}{2}\theta\right), \qquad \frac{\partial^2 \Phi}{\partial \theta^2} = -\left(\frac{N_p}{2}\right)^2 \Phi. \tag{1.6.21}$$

Equation (1.6.20) is satisfied when $\Phi \propto r^{\frac{N_p}{2}}$. The pseudopotential is proportional to r^{N_p-2}. The trap potential is given by the pseudopotential by the AC electric field and the DC electric field applied for the trapping in the z direction. Figure 1.12 shows the trap potential as a function of $r = \sqrt{x^2 + y^2}$, assuming that the DC electric field component in the xy direction is proportional to r.

1.7 Frequency of the AC electric field trap

There is a question about the proper frequency of the AC electric field trap. The AC volage supply is constructed of the power transfer with a coil with the conductance

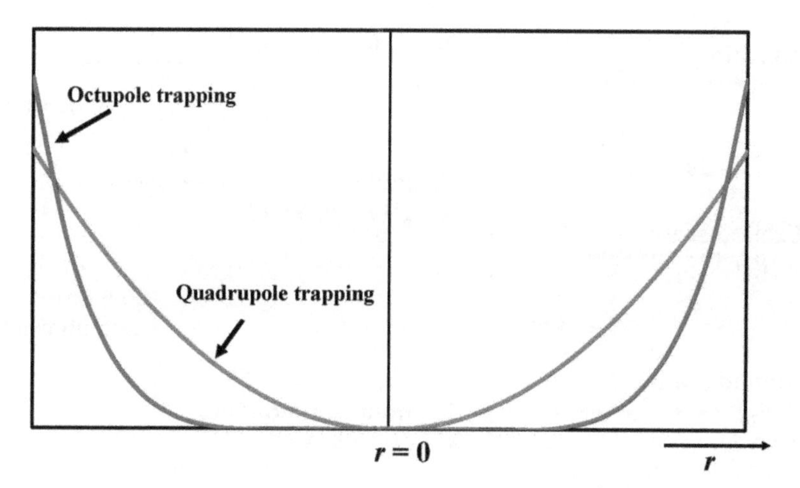

Figure 1.12. Trap potential using the quadrupole and octupole linear trap electrode, assuming that the expansion force by the DC electric field is proportional to r.

of L_c. The electrode has a capacitor of C_t as a condenser and the supply of high voltage is advantageous when

$$\Omega = \frac{1}{\sqrt{L_c C_t}}. \tag{1.7.1}$$

This frequency is higher for electrodes with smaller size (small capacity).

High frequency is required to satisfy equation (1.5.4) with a miniature electrode to trap a single or a few ions. The AC voltage with $\Omega = 2\pi \times (10\text{–}20)$ MHz is applied to an electrode with a gap smaller than 1 mm. Aimed at a trap of several hundred ions, AC voltage with $\Omega = 2\pi \times (100\text{–}500)$ kHz is applied to an electrode with the gap of a few cm.

1.8 Production of ions

The trapped ions are produced by the ionization of ions at the trap center. At first, ions are produced by the collision between an atomic (or molecular) beam and an electron beam from an electron gun (a stream of electrons from a hot cathode is focused and accelerated by an electric field). This method is applicable to the production of different kinds of ions, including rare gas ions [15].

Recently photoionization has been used more frequently to produce ions. Atomic or molecular transition to an excited state is induced by the first laser. The second laser ionizes the atoms or molecules in the excited state. For example, Ca^+ ions are produced by the procedure of (1) 4^1S_0–4^1P_1 transition of Ca atom by a laser light with the wavelength of 423 nm and (2) ionization by the excitation from the 4^1P_1 state by a laser light with the wavelength of 390 nm (see figure 1.13). The production efficiency of ions by photoionization is much higher than using an electron beam. We can perform an experiment without polluting of the surface of the electrode

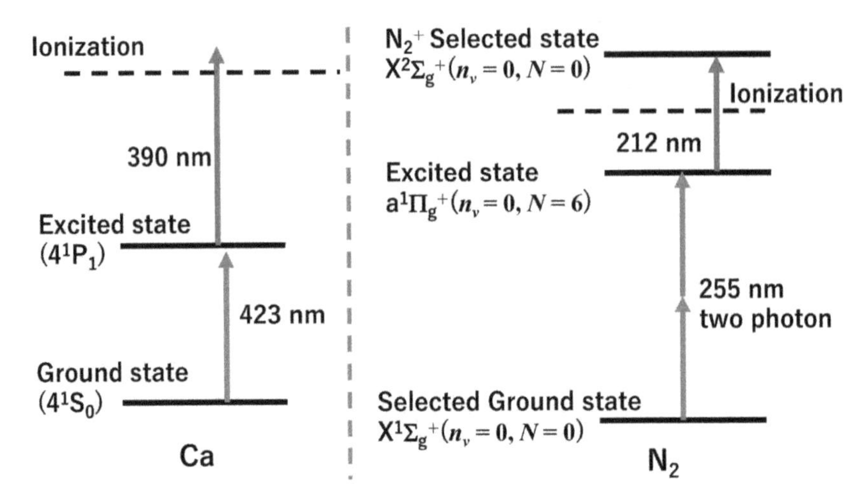

Figure 1.13. Photoionization of a Ca atom and a N_2 molecule with the description of states shown in section 2.1

(attachment of atoms). Using photoionization, we can produce ions of selected isotopes, because the transition frequencies depend on the isotopes.

Photoionization is also useful to prepare molecular ions in a selected vibrational–rotational state (see chapter 2). For example, a N_2^+ molecular ion was prepared in a selected vibrational–rotational state by the following procedure: (1) two-photon transition to an electronic excited state of N_2 molecules in a selected vibrational–rotational state is induced by the first laser light, and (2) N_2 molecules in the electronic excited state are ionized by the second laser light. Selecting the frequencies of both laser lights, produced N_2^+ molecular ions are localized in a selected vibrational–rotational state. Figure 1.13 shows the example to prepare N_2^+ molecular ions in the vibrational–rotational ground state [16].

References

[1] Kajita M 2019 *Measurement, Uncertainty and Lasers* (Bristol: IOP Publishing)
[2] Kajita M 2018 *Measuring Time: Frequency Measurement and Related Developments in Physics* (Bristol: IOP Publishing)
[3] Kajita M 2020 *Cold Atoms and Molecules* (Bristol: IOP Publishing)
[4] Ushijima I *et al* 2015 *Nat. Photon.* **9** 185
[5] Anderson M H, Ensher J R, Matthews M R, Wieman C E and Cornell E A 2014 *Science* **269** 198
[6] Department of Energy How particle accelerators work *Report*
[7] Huray P G 2010 *Maxwell's Equations* (New York: Wiley-IEEE) p 22
[8] Brown L S and Gabrielse G 1986 *Rev. Mod. Phys.* **58** 233
[9] Mathieu Stability Diagram for the Paul Trap (iisc.ac.in)
[10] Jones T 2006 *Mathieu's Equations and the Ideal rf-Paul Trap*
[11] Hoffges J *et al* 1997 *J. Mod. Phys.* **44** 1999
[12] Hoffges J *et al* 1997 *Opt. Commun.* **133** 170
[13] Keller M *et al* 2004 *Nature* **431** 1075

[14] Tanaka U *et al* 2009 *J. Phys. B: At. Mol. Opt. Phys.* **42** 154006
[15] Champenois C *et al* 2010 *Phys. Rev.* A **81** 043410
[16] Kajita M and Kimura N 2020 *J. Phys. B: At. Mol. Opt. Phys.* **53** 135401
[17] Germann M *et al* 2014 *Nat. Phys.* **10** 820

Chapter 2

Optical treatments of ions

In this chapter, the optical manipulations of ions are introduced. Atomic ions, constructed by nucleus and electrons, have discrete electron energy states. Vibrational–rotational energy states also exist for molecular ions. Ions absorb or emit electromagnetic wave (light, microwaves, etc) with the frequencies ν corresponding to the change of energy ΔE with the relation of $\Delta E = h\nu$ (h; Planck's constant). Using a cycle of light-induced excitation and the spontaneous emission deexcitation, we can also observe the fluorescence signal from a single trapped ion. This transition cycle is also used for optical pumping (localization to a single energy state) and laser cooling (reduction of kinetic energy). Ion crystal is formed when multiple ions are trapped, and laser cooled.

2.1 Energy structure of ions

The quantum mechanics shows that

(1) All particles have the characteristics of waves. On the other hand, electromagnetic waves have the characteristics of particles. The energy is given by $E = h\nu$ (h: Planck's constant, ν: frequency) and their momentum by $p = h/\lambda$ (λ: wavelength). As there is no significant change in the amplitude of the wavefunction within the phase difference of $\pm 1/2$ (phase has no physical appearance), there is an uncertainty principle between 'time and energy' and 'position and momentum'. When the particle is localized in an area of size L, the wavelength can only assume discrete values of $\lambda = 2L/(\text{integer})$, and hence also the discrete energy and momentum of the particles.

(2) The energy of the electrons in an atomic state can be changed ($E_1 \rightarrow E_2$) by absorbing or emitting a photon with an energy equal to the energy difference in the initial and final states (atomic transition energy). The total energy before and after the transition must be conserved. Therefore, atoms can only absorb or emit electromagnetic waves of a specific frequency satisfying $\nu_0 = |E_1 - E_2|/h$, known as the transition frequency or resonance

doi:10.1088/978-0-7503-5472-1ch2

frequency. The total momentum of atoms and photons before and after the transition must be also conserved in all directions.

(3) The lowest several energy states of atomic ions are mainly discussed with electron spin S (integer or half integer) and electron orbital angular momentum L (integer). There is also a quantum number of fine structure $J = L + S, L + S-1, \ldots, (L-S)$. The $J = L + S$ and $J = (L-S)$ states denote that the directions of the spin and orbital angular momentum are parallel and antiparallel, respectively. Using these quantum numbers, each energy state is described by $^{2S+1}[L]_J$, where $[L]$ is described by S, P, D, F for the $L = 0,1,2,3$ states, respectively. For example, $^2S_{1/2}$ denotes the $S = 1/2$, $L = 0$, $J = 1/2$ state.

There is also a quantum number $M_{S,\ L,\ J}$ as the component of S, L, and J in a direction and

$$M_J = M_S + M_L \tag{2.1.1}$$

is always satisfied. Each (J, M_J) state is given by

$^2S_{1/2}$ $\quad M_J = \pm\frac{1}{2} \quad (M_S, M_L) = \left(\pm\frac{1}{2}, 0\right)$

$^2P_{3/2}$ $\quad M_J = \pm\frac{3}{2} \quad (M_S, M_L) = \left(\pm\frac{1}{2}, \pm 1\right)$

$M_J = \pm\frac{1}{2}$ with zero magnetic field

$$(M_S, M_L) = \sqrt{\tfrac{2}{3}}\left(\pm\tfrac{1}{2}, 0\right) + \sqrt{\tfrac{1}{3}}\left(\mp\tfrac{1}{2}, \pm 1\right)$$

with high magnetic field:

$$(M_S, M_L) = \left(\pm\tfrac{1}{2}, 0\right)$$

$^2P_{1/2}$ $\quad M_J = \pm\frac{1}{2}$ with zero magnetic field

$$(M_S, M_L) = \sqrt{\tfrac{1}{3}}\left(\pm\tfrac{1}{2}, 0\right) - \sqrt{\tfrac{2}{3}}\left(\mp\tfrac{1}{2}, \pm 1\right)$$

with high magnetic field:

$$(M_S, M_L) = \left(\mp\tfrac{1}{2}, \pm 1\right)$$

$^2D_{5/2}$ $\quad M_J = \pm\frac{5}{2} \quad (M_S, M_L) = \left(\pm\frac{1}{2}, \pm 2\right)$

$M_J = \pm\frac{3}{2}$ with zero magnetic field

$$(M_S, M_L) = \sqrt{\tfrac{4}{5}}\left(\pm\tfrac{1}{2}, 1\right) + \sqrt{\tfrac{1}{5}}\left(\mp\tfrac{1}{2}, \pm 2\right)$$

with high magnetic field: $(M_S, M_L) = \left(\pm\frac{1}{2}, 1\right)$

$M_J = \pm\frac{1}{2}$ with zero magnetic field

$$(M_S, M_L) = \sqrt{\tfrac{3}{5}}\left(\pm\tfrac{1}{2}, 0\right) + \sqrt{\tfrac{2}{5}}\left(\mp\tfrac{1}{2}, \pm 1\right)$$

with high magnetic field:

$$(M_S, M_L) = \left(\pm\tfrac{1}{2}, 0\right)$$

$^2D_{5/2}$ $M_J = \pm\frac{3}{2}$ with zero magnetic field

$$(M_S, M_L) = \sqrt{\tfrac{1}{5}}\left(\pm\tfrac{1}{2}, 1\right) - \sqrt{\tfrac{4}{5}}\left(\mp\tfrac{1}{2}, \pm 2\right)$$

with high magnetic field:

$$(M_S, M_L) = \left(\mp\tfrac{1}{2}, \pm 2\right)$$

$M_J = \pm\frac{1}{2}$ with zero magnetic field

$$(M_S, M_L) = \sqrt{\frac{2}{5}}\left(\pm\frac{1}{2}, 0\right) - \sqrt{\frac{3}{5}}\left(\mp\frac{1}{2}, \pm 1\right)$$

with high magnetic field:

$$(M_S, M_L) = \left(\mp\frac{1}{2}, \pm 1\right). \tag{2.1.2}$$

The nucleus has also nuclear spin I (integer or half integer) and there is a hyperfine structure $F = J + I, J + I{-}1, \ldots, (J{-}I)$. The transition frequency between different hyperfine structure states is generally in the microwave region, while that between different fine structure states is in the optical or infrared region.

When there is an interaction with a laser light, the wavefunction at an energy eigenstate is mixed with those at other states and a change of state (transition) is caused, as shown in appendix A. The light indued transitions are mainly as follows:

E1 transition: interaction between electric field and the electric dipole moment

The transition is allowed between two states a and b with $\Delta J = 0, \pm 1$ (except for $J = 0$ -0). When the wavefunction of the state has a relationship with the transform of $r \rightarrow -r$ given by $\Phi(-r) = \Phi(r)$ (symmetric) or $\Phi(-r) = -\Phi(r)$ (anti-symmetric), the transitions between symmetric and anti-symmetric states are allowed. The transitions between two symmetric or two anti-symmetric states are forbidden. The transition rate is much higher than the E2 and M1 transitions, shown below. Forbidden a–b transitions are also possible when there is a mixture between the a and c states and the b–c transition is allowed. A mixture between the a and c states can be induced by an electromagnetic field. When the state mixture is induced by laser lights, the induced transition is called a 'multi-photon transition'. The two-photon transition is induced by two lasers with frequencies of ν_A and ν_B, when the transition frequency ν_0 is equal to $|\nu_A \pm \nu_B|$. The rate of two photon transition is high when ν_A and ν_B are close to the a–c and b–c transition frequencies, respectively. The transition with $\nu_0 = \nu_A + \nu_B$ is called 'two photon absorption'. Considering $\nu_A = \nu_B$, two photon absorption is also caused by one laser light with the frequency half of the transition frequency. The transition with $\nu_0 = | \nu_A - \nu_B |$ is called an 'induced Raman transition'. Two photon transition is allowed between two symmetric or two anti-symmetric states with $\Delta J = 0, \pm 1, \pm 2$. The $J = 0 \to 0$ transition also becomes possible by the mixture of the $J = 1$ state. Laser light with high intensity is required to induce the two photon transitions, therefore, the shift in the transition frequency induced by the light electric field (Stark shift shown in appendix B) is generally significant.

E2 transition: interaction between electric field gradient and the electric quadrupole moment.

The transition is allowed between two symmetric or two anti-symmetric states with $\Delta J = 0, \pm 1, \pm 2$(except for $J = 0 \to 0, 1/2 \to 1/2$). The rate of this transition is generally five orders smaller than that of the E1 transition.

M1 transition: interaction between the magnetic field and the magnetic dipole moment.

The transition is allowed between two symmetric or two anti-symmetric states with $\Delta J = 0, \pm 1$ (except for $J = 0 \to 0$). The transition rate is generally five orders smaller than the E1 transitions and same order with the E2 transitions.

For the light induced transitions, the $a \to b$ and $b \to a$ transition rates are equal. With the interaction of blackbody radiation, the population is derived to be uniform for all states, but the population at a lower energy state is higher than that in the higher energy state. This discrepancy is solved considering the transition from a higher energy state to a lower energy state, which is called 'spontaneous emission transition'. The spontaneous emission transition is caused without a trigger light, accompanied by the emission of light (fluorescence) with the transition frequency and a random phase and direction. The lifetime in an excited state is limited by the spontaneous emission transition rate. The phase of the wavefunction of the ion is jumped at the spontaneous emission transition, which makes the broadening of the transition spectrum by $\Gamma/2\pi$, where Γ is the rate of the spontaneous emission transition. The broadening of E1 allowed transition spectra in the optical region is generally larger than 1 MHz. The spontaneous emission transition is possible also with the E2 or M1 transitions, but their rates are generally lower than $2\pi \times 1$ kHz. 'Meta-stable states' are excited states with long lifetimes because there is no E1 allowed transition to a lower energy state.

The trap experiment has been performed using mainly alkali-like ions (Mg^+, Ca^+, Sr^+, Yb^+, Hg^+), whose energy structure is shown in figure 2.1. Table 2.1 lists the wavelength of laser light resonant with the $^2S_{1/2}$–$^2P_{1/2}$, $^2D_{3/2}$–$^2P_{1/2}$, and $^2S_{1/2}$–$^2D_{5/2}$ transitions [1].

For the isotopes of alkali-like ions with even mass numbers, the nuclear spin is zero and the energy structure is simple because there is no hyperfine structure. For isotopes with odd mass numbers, there is non-zero nuclear spin. Table 2.2 shows the transition frequencies between hyperfine structure states in the $^2S_{1/2}$ state [2].

The trap experiment has also been performed using alkali earth-like ions. The energy structure of alkali earth-like ions is shown in figure 2.2. With the alkali earth-like ions, $S = 0$ or 1 and the $^1S_0$ state is the ground state. The transition to the $^1P_1$ state is E1

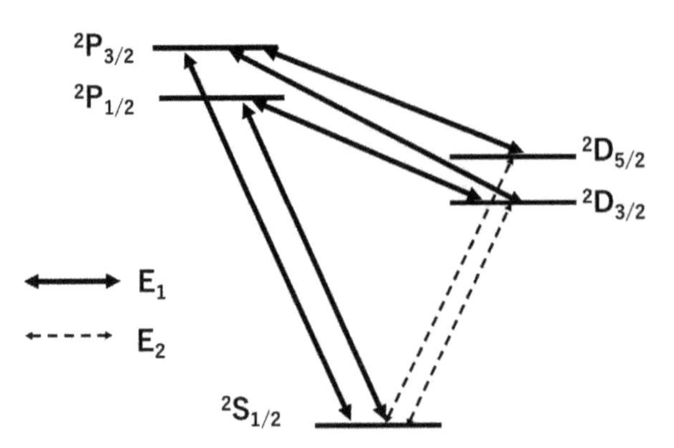

Figure 2.1. Energy structure of alkali-like ions. The E1 and E2 allowed transitions are shown with arrows.

Table 2.1. Wavelength of laser light resonant with the $^2S_{1/2}$–$^2P_{1/2}$, $^2D_{3/2}$–$^2P_{1/2}$, and $^2S_{1/2}$–$^2D_{5/2}$ transitions [1].

	$^2S_{1/2}$–$^2P_{1/2}$ (nm)	$^2D_{3/2}$–$^2P_{1/2}$ (nm)	$^2S_{1/2}$–$^2D_{5/2}$ (nm)
Be^+	313	–	–
Mg^+	280	–	–
Ca^+	397	866	729
Sr^+	422	1092	674
Ba^+	493	650	1760
Cd^+	214	–	–
Yb^+	369	935	411
Hg^+	193	–	282

Table 2.2. Nuclear spin, hyperfine states, and hyperfine transition frequencies of alkali-like ions with odd mass numbers [2].

	Nuclear spin	Hyperfine state	Transition frequency (GHz)
$^9Be^+$	3/2	$F = 1$–2	1.25
$^{25}Mg^+$	5/2	$F = 2$–3	1.79
$^{43}Ca^+$	7/2	$F = 3$–4	3.2
$^{87}Sr^+$	9/2	$F = 4$–5	5.0
$^{135}Ba^+$	3/2	$F = 1$–2	7.2
$^{113}Cd^+$	1/2	$F = 0$–1	15.2
$^{171}Yb^+$	1/2	$F = 0$–1	12.6
$^{199}Hg^+$	1/2	$F = 0$–1	40.5

allowed. The electron spin changing transition is forbidden, therefore, ^3P states (lowest from the $S = 1$ states) are meta-stable. However, there is a mixture between $^1P_1$ and $^3P_1$ states by the LS-coupling and the $^1S_0$–$^3P_1$ transition is possible with the rate almost two orders smaller than that for the $^1S_0$–$^1P_1$ transition. The $^1S_0$–$^3P_0$ transition is possible

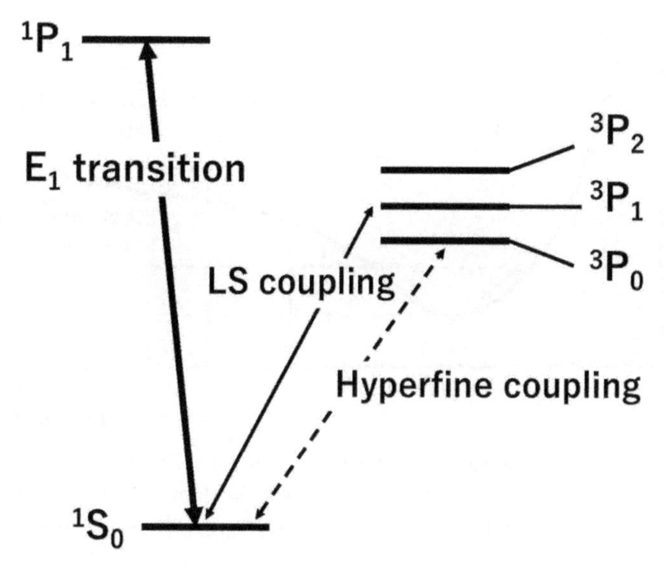

Figure 2.2. Energy structure of alkali earth-like ions.

Table 2.3. Wavelength of laser light resonant with the $^1S_0$–$^1P_1$, $^1S_0$–$^3P_1$, and $^1S_0$–$^3P_0$ transitions [2–5].

	$^1S_0$–$^1P_1$ (nm)	$^1S_0$–$^3P_1$ (nm)	$^1S_0$–$^3P_0$ (nm)
$^{27}Al^+$	167	267.0	267.4
$^{119}In^+$	159	230	237

only when a coupling (a mixture of eigen wavefunctions, see section 3.1) between the $^3P_0$ and $^3P_1$ states induced by the magnetic field. When there is non-zero nuclear spin, it works like a permanent magnet and this transition has been observed with $^{27}Al^+$ and $^{119}In^+$ ions [2–5]. The wavelength of laser light resonant with the $^1S_0$–$^1P_1$, $^1S_0$–$^3P_1$, and $^1S_0$–$^3P_0$ transitions are listed in table 2.3 [2–5].

The energy structure of molecular ions is much more complicated, because of the relative motion between atoms, which construct the molecular ion. Here we consider just the diatomic molecular ions.

It is difficult to analyze the energy structure of molecules, taking the motion of all nuclei and electrons into account. The Born–Oppenheimer approximation [6] was proposed to calculate the energy of electrons in a molecule E_e taking constant values of the internuclear distance r. The dependence of E_e on r gives a potential energy curve, as shown in figure 2.3. This approximation is based on the idea that the motion of the nucleus is much slower than that of the electron, because the mass of the nucleus is much larger than that of the electron. At the bonding electronic states, E_e is minimum with a certain value of r, called the bond length r_b, as shown in figure 2.3. E_e is generally displayed by the value at $r = r_b$, because the nucleus

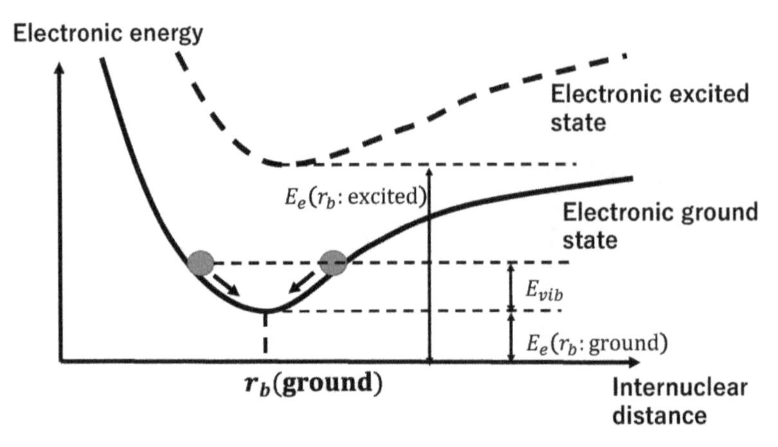

Figure 2.3. Electronic energy E_e of diatomic molecules as a function of internuclear distance r. For bonding states, E_e is minimum at a certain distance called the bonding length r_b. The electronic energy is generally defined by $E_e(r_b)$, as the nucleus vibrates at around $r = r_b$. $E_e(r) - E_e(r_b)$ is considered to be the vibrational energy E_{vib}.

oscillates around $r = r_b$ and $E_e(r) - E_e(r_b)$ is interpreted as the vibrational motion energy (shown below), as shown in figure 2.3. For the diatomic molecular ion, electron orbital angular momentum is defined only by the component parallel to the molecular axis K_L. The electronic ground states are described by $X^{2S+1}[K_L]$, where $[K_L]$ is described as Σ, Π, and Δ with $K_L = 0,1$, and 2, respectively. S is the electron spin. In this book, we discuss diatomic molecular ions in the $X^1\Sigma$ and $X^2\Sigma$ states except for that the O_2^+ molecular ion in the $X^2\Pi$ state is mentioned in section 5.6.4.

The vibrational motion is the relative motion between the bounded nucleus parallel to the bonding direction (energy E_{vib}). The relative motion perpendicular to the bonding direction is the rotational motion (energy E_{rot}).

The potential energy curve is approximated by a harmonic potential at around r_b (low vibrational state). The harmonic vibrational energy is given by

$$E_{\text{vib}} = \left(n_v + \frac{1}{2}\right)h\nu_{\text{vib}} \tag{2.1.3}$$

The $n_v = 0 \to 1$ transition frequency is roughly ν_{vib}. The population in the $n_v = 1$ state is generally two or three orders smaller than that in the $n_v = 0$ state.

The rotation of molecules in the $^{2S+1}\Sigma$ state is described by the rotational angular momentum quantum numbers N and M_N (integer $-N \leqslant M_N \leqslant N$). The angular momentum component in one direction is given by $(h/2\pi)M_N$ and the square of the absolute value of angular momentum is $(h/2\pi)^2 N(N + 1)$. Then the rotational energy is given by

$$E_{\text{rot}} = hB_{n_v}N(N + 1)$$

$$B_{n_v} = \frac{h}{8\pi^2 I_{n_v}} I_{n_v}\text{: moment of inertia.} \tag{2.1.4}$$

B_{n_v} is the rotational constant at each vibrational state. The $N \to N + 1$ rotational transition frequency is $\nu_{rot} = 2B_{n_v}(N + 1)$. The moment of inertia in the vibrational ground state is approximately given by $I_0 = \mu_r r_b^2$, where μ_r is the reduced mass (see appendix C). With a higher vibrational state, I_{n_v} is larger than I_0 (B_{n_v} is smaller than B_0) because the average of $(r_b \pm \delta r)^2$ is $r_b^2 + (\delta r)^2$ (δr is the amplitude of the vibration).

The population is maximum with $N = 10$–20 for many molecules with the room temperature ($T = 300$ K).

For the diatomic molecules in the $^{2S+1}\Sigma$ ($S \geqslant 1/2$) state, there is an energy splitting $E_{spin\text{-}rot}$ by the spin-rotation coupling with a quantum number J ($=N + S, N + S - 1, \ldots, |N - S|$). In the $^1\Sigma$ state, the rotational state is described by J, because of $J = N$. There is also hyperfine structure E_{hf} when there are non-zero nuclear spins with forming atoms. Then the molecular energy state is given by

$$E_{mol} = E_e(r = r_b) + E_{vib} + E_{rot} + E_{spin-rot} + E_{hf} \tag{2.1.5}$$

There is a special quantum characteristic with homonuclear diatomic molecules. The states of relative position $\vec{r}$ and $-\vec{r}$ are not distinguishable and there must be an interference between both states like

$$\Phi_{\pm} = \frac{\Phi(\vec{r}) \pm \Phi(-\vec{r})}{\sqrt{2}} \tag{2.1.6}$$

Particles with integer spin called bosons have positive interference Φ_+, and particles with half integer particles called fermions have negative interference Φ_-. The wavefunction of an even N state is symmetric (no change of sign) and that of an odd N state is anti-symmetric (change the sign) by the exchange of the position. When there is nuclear spin i_{nuc}, the total spin state of $I_{nuc} = 2i_{nuc} - k_{nuc}(0 \leqslant k_{nuc} \leqslant 2i_{nuc})$ is symmetric with even k_{nuc} and anti-symmetric with odd k_{nuc} with the exchange of the position. For homonuclear diatomic molecules with boson (fermion) nuclei, only states with even (odd) $N + k_{nuc}$ can exist. For example, the H nucleus is fermion. Therefore, only even rotational state can exist with $I_{nuc} = 0$ (para) state and only odd rotational states can exist with $I_{nuc} = 1$ (ortho) state.

2.2 Optical pumping

The population in the different quantum states in atoms or molecules (neutral or ionic) are of the same order when the energy difference between them is smaller than $k_B T$ (k_B: Boltzmann constant, T: thermal dynamic temperature). Optical pumping is a method to localize them in a single state. Here we consider two ground states $g_{1,2}$ and one excited state e. When the $g_2 \to e$ transition is excited by a laser light, the deexcitation is caused by the spontaneous emission transitions to the g_1 or g_2 state. The transition from the g_2 state to the g_1 state is possible, while no transition is induced from the g_1 state. The population is finally totally localized to the g_1 state, as shown in figure 2.4. This procedure is called 'optical pumping'. No fluorescence is observed after the optical pumping.

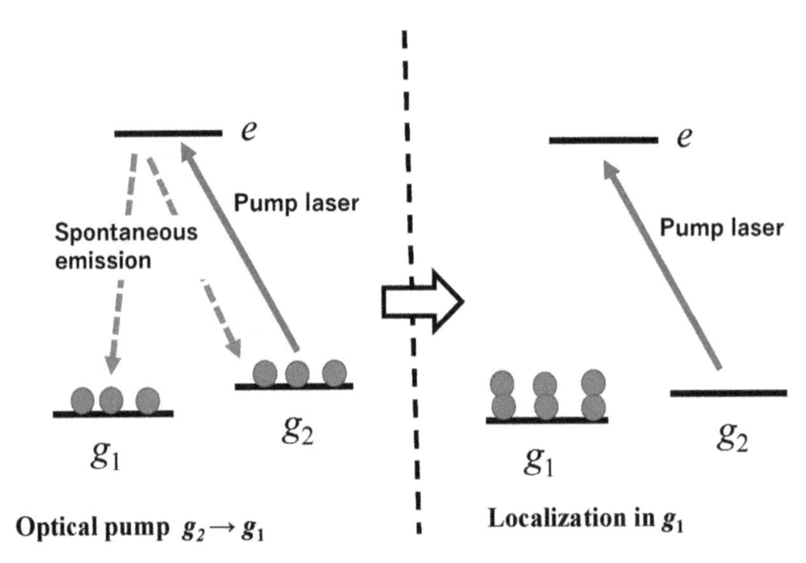

Figure 2.4. The procedure of optical pumping with the energy structure of $g_{1,2}$ and e states. With a laser light, which is resonant to the g_2–e transition, the distribution is localized to the g_1 state.

For example, we consider the even isotopes of alkali-like ions are irradiated by a laser light which is resonant to the $^2S_{1/2}$–$^2P_{1/2}$ transition. In the $^2S_{1/2}$ state, there are two substates of $M_J = \pm 1/2$. When the light is circularly polarized in the right (σ^+) direction, only the $\Delta M_J = 1$ transition is induced. When σ^+ polarized light was irradiated, the $^2S_{1/2}$ $M_J = -1/2 \rightarrow {}^2P_{1/2} \rightarrow {}^2S_{1/2}$ $M_J = 1/2$ or the $^2S_{1/2}$ $M_J = -1/2 \rightarrow {}^2P_{1/2} \rightarrow {}^2D_{3/2} \rightarrow {}^2S_{1/2}$ $M_J = 1/2$ transitions are induced, while no transition is induced from the $M_J = 1/2$ state. The population is finally localized in the $^2S_{1/2}$ $M_J = 1/2$ state.

When the ground state is given by more than two states, the distribution is localized to one state g_1 when laser light induces excitation from all states except for the g_1 state.

2.3 Monitoring the quantum state by the quantum jump

Here we consider the system of three states: ground state g, excited state e, and intermediate state m. The g–e and m–e transitions are E1 allowed, while g–m transition is E1 forbidden. From the e state, the probability of the spontaneous emission transition to the m state is assumed to be much smaller than that to the g state. This assumption is generally valid because the spontaneous emission rate is proportional to the cubic of the transition frequency.

A laser light with the resonant frequency of the g–e transition is irradiated. When the ion is in the g state, the fluorescence from the $e \rightarrow g$ spontaneous emission is observed. When the ion is in the m state, no fluorescence is observed. Observing the fluorescence from many trapped ions, the intensity of the fluorescence signal is almost constant, because the population ratio between the g and m states is almost constant.

If we observe the fluorescence from a single trapped ion, it is observed as a signal of the dark (no fluorescence) and bright (fluorescence) states, as shown in figure 2.5.

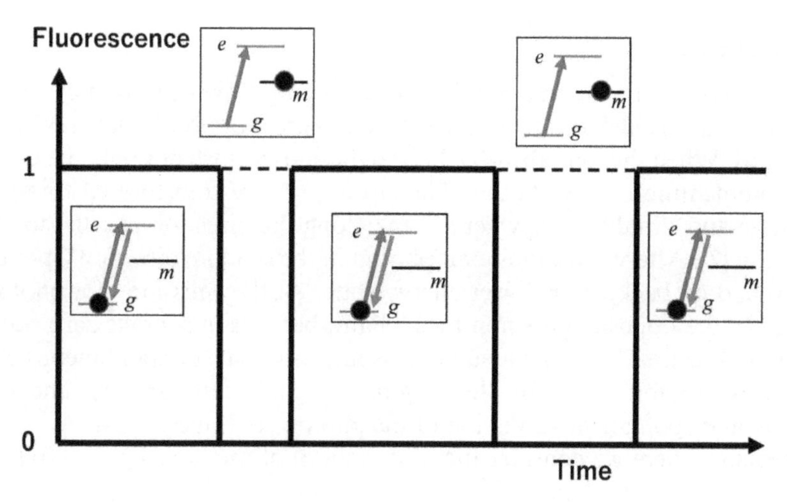

Figure 2.5. The observation of quantum jump. We consider the energy states of a single ion with the ground state *g*, excited state *e*, and medium state *m*. With a laser light resonant to the *g*–*e* transition, we observe the fluorescence when the ion is in the *g* state. When the ion is in the *m* state, no fluorescence is observed.

The dark state indicates the ion being in the *m* state, while the bright state indicates the *g* state. The dark state continues until the $m \rightarrow g$ spontaneous emission transition (for example, by E2 transition) is caused. The change between the bright and dark states is called 'quantum jump'. The mean period of the dark state indicates the lifetime in the *m* state.

When a single alkali-like ion is trapped and a laser light with a frequency resonant to the $^2S_{1/2}$–$^2P_{1/2}$ transition, the fluorescence is observed while the ion is in the $^2S_{1/2}$ state (bright state). Once the $^2P_{1/2}$–$^2D_{3/2}$ spontaneous emission transition is caused, we cannot observe the fluorescence while the ion remains in the $^2D_{3/2}$ state. The quantum jump indicates whether the ion in the $^2S_{1/2}$ or $^2D_{3/2}$ state.

The quantum jump gave an interpretation to the quantum mechanics that 'the quantum energy state converges to one eigenstate with the treatment to determine the energy state (section 3.1)'.

2.4 Laser cooling of trapped ions

Laser cooling is a technology to reduce the amplitude of the vibrational motion of trapped ions. For the RF trap, the ion motion is given by the secular motion and the micromotion, as shown in section 1.5. The amplitude of the secular motion is reduced by laser cooling. The amplitude of the micromotion is not reduced by laser cooling, as the amplitude of micromotion is rigorously given by the amplitude of the RF electric field. The loss of micromotion energy by laser cooling is compensated by the energy supplied from the RF electric field. It is important to adjust the position of the minimum trap potential to the position where the RF electric field is zero (see section 1.5). With this condition, the amplitude of the micromotion is also reduced as the amplitude of the secular motion is reduced.

2.4.1 Doppler cooling

At first, we discuss the Doppler cooling assuming that the transition cycle between the ground and excited states is much faster than the vibrational motion of the trapped ions. When the ions absorb a laser light, ions absorb not only the energy but also the momentum of the photon. Therefore, ions are transformed to an excited energy state and simultaneously get a force along the direction parallel to the light propagation [7]. Afterwards, ions emit photon in the random direction (spontaneous emission) and go back to the lower energy state. At the emission of a photon, ions are forced in the opposite direction to the light, but it is in average zero because of the random direction. Here we assume for simplicity that the spontaneous emission transition is possible only for the original ground state, so that the cycle of absorption and spontaneous emission of the photon continues. Then ions are forced in the direction (here we consider the x -direction) of the laser light as follows

$$F_x = \frac{h\nu}{c}R(\nu)\nu\text{: laser light frequency,} \tag{2.4.1}$$

where $R(\nu)$ is the rate of the cycle of laser induced excitation and the spontaneous emission deexcitation given by

$$R(\nu) = \frac{\Gamma\Omega_R^2}{\Delta^2 + (\Gamma/2\pi)^2}, \qquad \Delta = \nu_0 - \nu$$

ν_0: transition frequency Ω_R: Rabi frequency (appendix A)

Γ: spontaneous emission rate. (2.4.2)

The change of secular motion energy (kinetic energy + pseudo potential) of ion E_V with the velocity of $v_x = \sqrt{2E_V/m}\sin(2\pi v_m t)$ is given by

$$\frac{dE_v}{dt} = mv_x\frac{dv_x}{dt} = v_x F_x = \frac{h\nu}{c}v_x R\left(\nu\left(1 + \frac{v_x}{c}\right)\right)$$

$$R\left(\nu\left(1 + \frac{v_x}{c}\right)\right) \approx R(\nu)\left(1 + \frac{2\Delta\nu(v_x/c)}{\Delta^2 + (\Gamma/2\pi)^2}\right). \tag{2.4.3}$$

The time average of dE_V/dt is given by

$$\langle v_x R(\nu)\rangle_{\text{ave}} = 0, \qquad \langle v_x^2\rangle_{\text{ave}} = \frac{E_V}{m}$$

$$\left\langle\frac{dE_v}{dt}\right\rangle_{\text{ave}} = R(\nu)\frac{h\nu}{c}\frac{2\Delta\nu}{\Delta^2 + (\Gamma/2\pi)^2}\frac{E_V}{mc} = R(\nu)\frac{h\nu^2}{mc^2}\frac{2\Delta E_V}{\Delta^2 + (\Gamma/2\pi)^2}. \tag{2.4.4}$$

Equation (2.4.4) indicates $dE_V/dt < 0$ when $\Delta < 0$ (red detuned). This result shows that the deceleration force (while $v_x < 0$) is higher than the acceleration force (while $v_x > 0$) and the kinetic energy of the trapped ions are reduced, as shown in figure 2.6. When $\Delta > 0$ (blue detuned), it gives a heating effect. Figure 2.7 indicates the image of the fluorescence signal when a laser light is irradiated to the trapped ions. Because of the cooling effect by a red detuned laser and the heating effect by a blue detuned laser, the dependence of the observed fluorescence intensity on the laser frequency is not symmetric to the transition frequency.

Laser cooling of trapped ions can be performed using one laser in the direction with non-zero components in x,y,z-directions. Therefore, the cooling apparatus for trapped ions is much simpler [8] than that for the neutral atoms in a free space, for which red detuned laser lights should be irradiated from two (six) directions for the one-(three-)dimensional cooling [9].

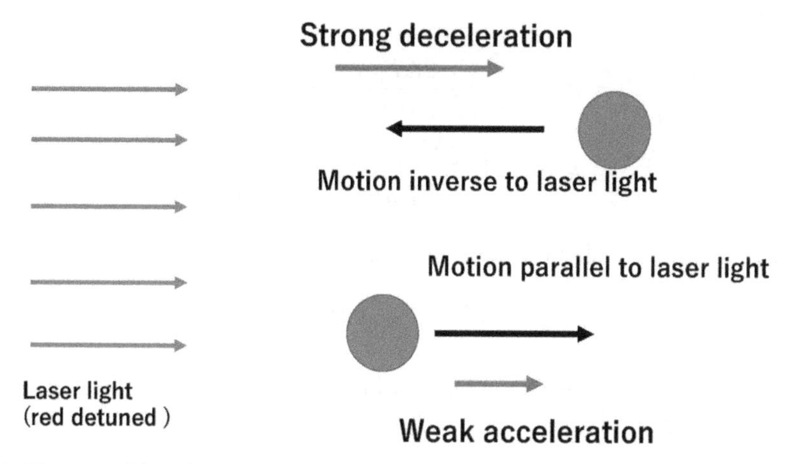

Figure 2.6. The trapped ions have vibrational motion and get a deceleration (acceleration) force when the motion direction is inverse (parallel) to the direction of the laser light. When the frequency of the laser light is lower than the transition frequency (red detuned), the deceleration force is stronger than the acceleration force.

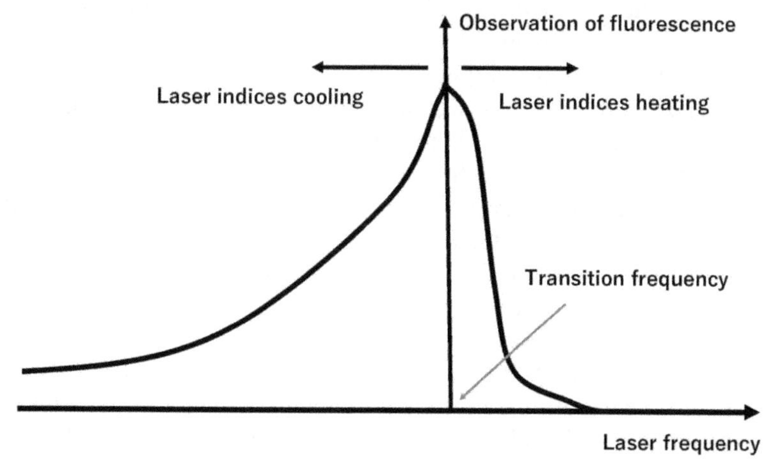

Figure 2.7. The image of the fluorescence signal when a laser light is irradiated to the trapped ions. The dependence of the observed fluorescence intensity to the laser frequency is not symmetric to the transition frequency because the red (blue) detuned laser gives a cooling (heating) effect.

As the ions are decelerated and the first order Doppler shift becomes smaller than the transition line width $\Gamma/2\pi$ (given by the limited interaction time by spontaneous emission), the difference between interaction forces at parallel and anti-parallel motion to the laser light becomes smaller. There is also a heating effect caused by the momentum of photons emitted in the random direction at the spontaneous emission. The heating rate is given by

$$E_{\text{scat}} = \frac{1}{2m}\left(\frac{h\nu}{c}\right)^2 R(\nu). \tag{2.4.5}$$

Considering the change of the secular motion energy also taking the heating effect into account

$$\begin{aligned}\left\langle \frac{dE_v}{dt} \right\rangle_{\text{ave}} &= R(\nu)\frac{h\nu^2}{mc^2}\frac{2\Delta E_V}{\Delta^2 + (\Gamma/4\pi)^2} - \frac{1}{2m}\left(\frac{h\nu}{c}\right)^2 R(\nu) \\ &= R(\nu)\frac{1}{2m}\left(\frac{h\nu}{c}\right)^2\left[\frac{4\Delta E_V}{h\{\Delta^2 + (\Gamma/2\pi)^2\}} - 1\right],\end{aligned} \tag{2.4.6}$$

and the equilibrium secular motion energy is given by

$$E_{V-\text{eq}} = \frac{h}{4\Delta}\{\Delta^2 + (\Gamma/2\pi)^2\}, \tag{2.4.7}$$

which is minimum with $\Delta = \Gamma/2\pi$. The minimum equilibrium secular motion energy (called the Doppler limit)

$$E_{\text{Doppler}} = \frac{h\Gamma}{4\pi}. \tag{2.4.8}$$

But the attainable kinetic energy of RF trapped ion is in reality higher (order of 1 mK) than the Doppler limit because of the heating effect by the transform from the micromotion energy at collision between trapped ions (see section 1.5).

The above discussion assumes that the spontaneous emission transition from an excited state e is possible only to one ground state g. When the spontaneous emission transition from the state e is possible to two states $g_{1,2}$, laser cooling is not possible with one laser light. When laser cooling is performed with the g_1–e transition, the ion is pumped to the g_2 state after several times of cooling cycle (section 2.2) and there is no cooling effect afterwards. Given an additional laser resonant to the g_2–e transition (repump laser), the cooling continues. Note that both (cooling + repump) transitions are suppressed when the detuning of both lasers from each transition frequencies are equal. This phenomenon is called electric induced transparency (EIT) shown in appendix A. With the EIT state, there is no interaction with the laser light and the quantum state called coherent population trapping (CPT) remains [10].

Doppler cooling using E1 allowed transition is useful to reduce the kinetic energy rapidly, but several other cooling methods have been developed to attain the kinetic energy below the Doppler limit as shown below.

2.4.2 Sideband cooling

Cooling using E1 forbidden (for example E2) transitions seems to be useful, because the spectrum linewidth is narrow, and the transition rate is sensitive to the slight difference of the kinetic energy of trapped ion. The E1 forbidden transition is slower than the vibrational motion, therefore, the Doppler effect is considered with the averaged effect within the period of the vibrational motion (temporal effect for E1 allowed transition). For the transition frequency ν_0, the Doppler effect of the vibrational motion with $r = r_0 \sin(2\pi\nu_m t)$ is given by [11]

$$\sin\left[2\pi\nu_0\left\{1 + \frac{r_0\nu_m}{c}\cos(2\pi\nu_m t)\right\}t\right] = \sin\left[2\pi\nu_0 t + \frac{2\pi r_0\nu_m\nu_0}{c}\int\cos(2\pi\nu_m t)dt\right]$$

$$= \sin\left[2\pi\nu_0 t + \frac{r_0}{\lambda_L}\sin(2\pi\nu_m t)\right] = \sum_{n_s=-\infty}^{\infty} J_{n_s}\left(\frac{r_0}{\lambda_L}\right)\sin[2\pi(\nu_0 + n_s\nu_m)t + \theta_{n_{\text{vib}}}]$$

$\lambda_L = \frac{\nu_0}{c}$ (wavelength of laser light), n_s: integer

$$J_{n_s}: n_s - \text{th order Bessel function } \theta_{n_{\text{vib}}}: \text{proper phase shift.} \tag{2.4.9}$$

Equation (2.4.9) shows that the transition frequency components are given by $\nu_0 \pm n_s\nu_m$. When $r_0 \ll \lambda_L$ (Lamb–Dicke regime), the components of $|n_s| \geqslant 1$ are negligibly small. We can give a simple interpretation of equation (2.4.9) using the first and second terms of Taylor expansion

$$\begin{aligned}\sin\left[2\pi\nu_0 t + \frac{r_0}{\lambda_L}\sin(2\pi\nu_m t)\right] &= \sin[2\pi\nu_0 t] + \frac{r_0}{\lambda_L}\cos(2\pi\nu_0 t)\sin(2\pi\nu_m t)\\ &= \sin[2\pi\nu_0 t] + \frac{r_0}{2\lambda_L}\sin[2\pi(\nu_0 + \nu_m)t]\\ &- \frac{r_0}{2\lambda_L}\sin[2\pi(\nu_0 - \nu_m)t].\end{aligned} \tag{2.4.10}$$

The the trapped ion is given vibrational motion energy by $(n_{\text{vib}} + 1/2)h\nu_m$. When the transition is induced by the laser light with the frequency of $\nu = \nu_0 + n_s\nu_m$, the motion vibrational quantum number changes by $n_{\text{vib}} \to n_{\text{vib}} + n_s$. We call this the red sideband transition for $n_s < 0$ and blue sideband transition for $n_s > 0$. Sideband cooling is a method to reduce the vibrational motion energy with the procedure of (1) $(g, n_{\text{vib}}) \to (e, n_{\text{vib}} - 1)$ transition is induced using a laser with the frequency of $\nu = \nu_0 - \nu_m$ and (2) $(e, n_{\text{vib}} - 1) \to (g, n_{\text{vib}} - 1)$ spontaneous emission transition. Figure 2.8 shows the procedure to reduce $n_{\text{vib}} = 2$ to $n_{\text{vib}} = 0$. With this method, the vibrational motion energy is reduced to $n_{\text{vib}} = 0$ with the energy of $h\nu_m/2$. The secular motion of a trapped ion is given by several motion frequencies ν_m^α (α = 1,2, ...,) as shown in appendix C. With sideband cooling, $\nu = \nu_0 - \nu_m$ is satisfied only with one ν_m^α. To perform sideband cooling with all motion frequency components, it should be performed changing ν_m between different values of ν_m^α. This method is not useful when $\Gamma/2\pi > \nu_m$, because the tail of the blue sideband component overlaps at

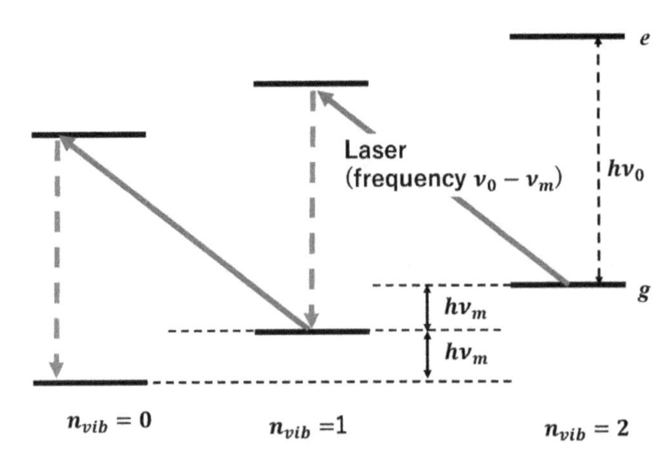

Figure 2.8. Procedure of sideband cooling to reduce n_{vib} from 2 down to 0.

$\nu = \nu_0 - \nu_m$. The sideband cooling should be performed after the Doppler cooling using a E1 allowed transition so that the motion energy is reduced to the order of 1 mK.

2.4.3 Sideband Raman cooling

Sideband Raman cooling is useful when there are two ground states $g_{1,2}$, which is performed with the following procedure [11].

(1) The $(g_1, n_{\text{vib}}) \rightarrow (g_2, n_{\text{vib}} - 1)$ transition is induced using Raman lasers 1 and 2. The frequencies of the Raman lasers 1 and 2 are $\nu_{R1} = \nu_{01} - \delta_R$ and $\nu_{R2} = \nu_{02} + \nu_m - \delta_R$, respectively ($\nu_{0,1}$: the $g_{1,2}$–e transition frequency, δ_R: arbitral detuning). The CPT state is constructed with the population ratio in both states, which is determined by the intensity ratio of two Raman lasers. Reducing the intensity of Raman laser 2 adiabatically to zero, the one-way transition is performed (see appendix A).

(2) Using a laser with a frequency of ν_{02}, the optical pumping is performed with the $(g_2, n_{\text{vib}} - 1) \rightarrow (e, n_{\text{vib}} - 1) \rightarrow (g_1, n_{\text{vib}} - 1)$ to the state is performed.

The (1) and (2) procedures are repeated as shown in figure 2.9, until $n_{\text{vib}} = 0$ is attained.

2.4.4 EIT cooling [12]

EIT cooling is a method in which the cooling procedure is much faster than the sideband and sideband Raman cooling. With this method, red sideband transitions ($n_s < 0$) are induced, while the carrier ($n_s = 0$) and blue ($n_s > 0$) sideband transitions are suppressed by EIT effect (appendix A). The cooling is performed with a cooling laser with a frequency of $\nu_{\text{cool}} = \nu_{01} + \Delta_{\text{cool}}$ and a coupling laser with a frequency of $\nu_{\text{couple}} = \nu_{02} + \Delta_{\text{couple}}$ ($\nu_{01,02}$: the $g_{1,2} - e$ transition frequency) (see figure 2.10). The directions of both lasers are counter propagating so that the red sideband transition rate is high enough for cooling. The $g_{1,2} - e$ transitions are E1 allowed. The intensity

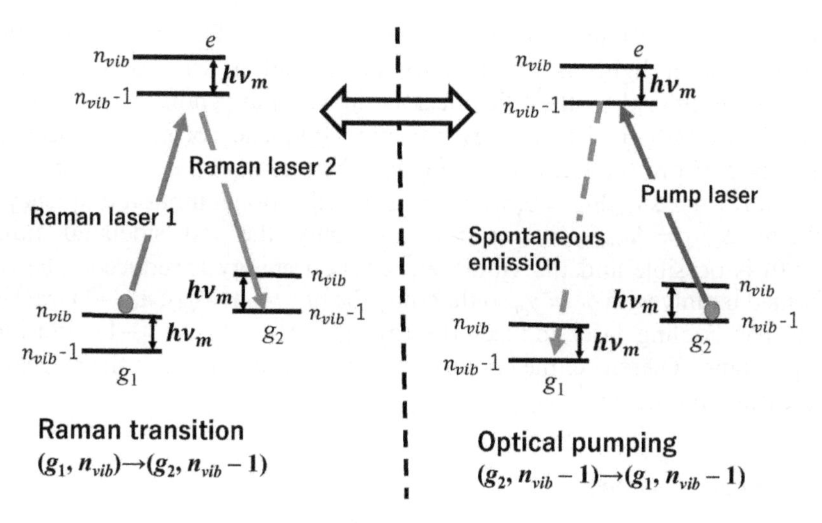

Figure 2.9. Procedure of sideband Raman cooling.

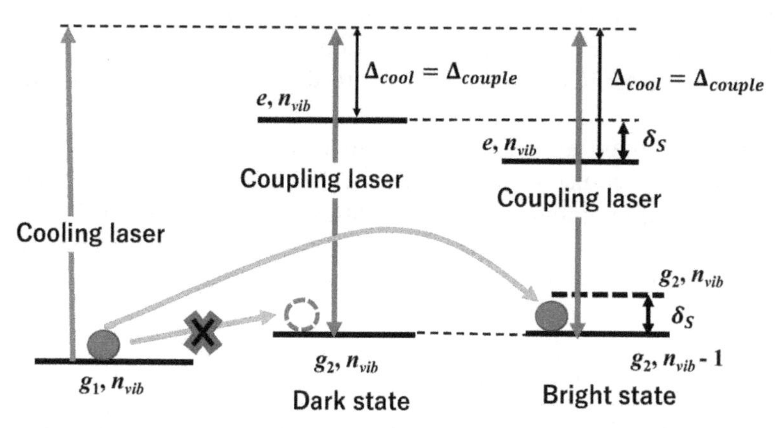

Figure 2.10. The fundamental of EIT cooling. The directions of cooling and coupling lasers are counter propagating. The coupling laser does not induce any transition from the g_1 state. The frequency detuning of the cooling laser to the $g_1 \to e$ transition Δ_{cool} and that of the coupling laser to the $g_2 \to e$ transition Δ_{couple} are equal. The g_1 state is the coupling of dark and bright states. With the dark state, there is no Stark shift in the Raman transition frequency. The frequency difference between two laser lights is resonant to the $(g_1, n_{\text{vib}}) \to (g_2, n_{\text{vib}})$ transition, but this transition is suppressed by the EIT effect. With the bright state, there is coupling with the excited (e) state and there is Stark shift δ_S in the transition frequency. When $\delta_S = \nu_m > 0$, only the $(g_1, n_{\text{vib}}) \to (g_2, n_{\text{vib}} - 1)$ transition is possible, where ν_m is the frequency of the vibrational motion of the ion.

of the coupling laser is high enough to induce a significant Stark energy shift in the g_2 (δ_S) and e ($-\delta_S$) states (appendix B), but it does not induce the coupling between the g_1 and e states because of its polarization.

We consider the possibility of the $(g_1, n_{\text{vib}}) \to (g_2, n_{\text{vib}} + \Delta n_{\text{vib}})$ Raman transition with $\Delta_{\text{cool}} = \Delta_{\text{couple}}$, where n_{vib} is the quantum number of the motion vibrational state. When the ion is pumped to the g_1 state, it is a couple of the dark (EIT) state and bright state. For the dark state, there is no interaction with the e state, therefore, there is no Stark shift induced by the coupling laser. The frequency difference between both laser lights is resonant to the transition with $\Delta n_{\rm vib} = 0$, but this transition is suppressed by the EIT effect as shown in appendix A. With the bright state, the Stark shift in the $g_1 \rightarrow g_2$ transition frequency of δ_S is induced by the coupling laser. From the frequency difference between two lasers, the transition is induced with $\delta_S + (\Delta n_{\rm vib})\nu_m = 0$, where ν_m is the vibrational motion frequency of the ion. When $\Delta_{\rm cool} = \Delta_{\rm couple} > 0$, $\delta_S > 0$ and only the red sideband transition ($\Delta n_{\rm vib} < 0$) is possible and the vibrational motion energy is reduced. The cooling is performed mainly with $\delta_S = \nu_m$ so that only the $(g_1, n_{\rm vib}) \rightarrow (g_2, n_{\rm vib}-1)$ transition is induced. The cooling laser induces the $(g_2, n_{\rm vib}-1) \rightarrow (g_1, n_{\rm vib}-1)$ transition by optical pumping. Therefore, the cooling procedure is simpler than the sideband and sideband Raman cooling.

2.4.5 Sympathetic cooling

As shown above, laser cooling is possible with ions with simple energy structure, for example alkali-like ions. Molecular ions are difficult to laser cool, because of the complicated energy structure with vibrational rotational states. Laser cooling of alkali earth-like ions using the $^1S_0$–$^1P_1$ transition is in principle possible, but it is not realistic because a laser light with a wavelength shorter than 180 nm is required (see table 2.3).

Ions with any quantum energy structure can be cooled with the collisional interaction with laser cooled ions. We consider the change of the kinetic energy when ion 1 with a mass of m_1 collides with ion 2 with a mass of m_2. The velocity of ions 1 and 2 are assumed to be v_i and 0, respectively. After the elastic collision, the velocity of ion 1 is $(m_1 - m_2)v_i/(m_1 + m_2)$ and the change of the kinetic energy is given by

$$\Delta K_{\rm ion} = -\frac{4m_1m_2}{(m_1 + m_2)^2}K_{\rm ion}, \qquad K_{\rm ion} = \frac{m_1v_i^2}{2}. \tag{2.4.11}$$

Equation (2.4.11) shows that the sympathetic cooling effect is efficient when the masses of both ions are close.

Using sympathetic cooling, ultra-low kinetic energy can be obtained with different kinds of ions, including molecular ions [13] and alkali earth-like ions [3–5].

2.5 Crystalization of laser cooled ions

A single trapped ion has a secular vibrational motion. Its amplitude is reduced by laser cooling, and it is localized at the position where the pseudo potential is minimum. We consider this position as the origin of the coordinate ($\vec{r} = 0$).

When multiple ions are trapped, there is a repulsive Coulomb force between ions. For simplicity, we consider when two ions with same mass and electric charge of $+e$ are trapped; when one ion is localized at $\vec{r}$, another ion is localized at $-\vec{r}$ assuming that the trap potential energy distribution is symmetric $\left(P_p(\vec{r}) = P_p(-\vec{r})\right)$. The trapped ions are localized at $\pm\vec{r}_c$, where the total potential energy given by

$$P_{\text{tot}} = \frac{e^2}{8\pi\varepsilon_0 \mid \vec{r} \mid} + P_p(\vec{r}) + P_p(-\vec{r}) = \frac{e^2}{8\pi\varepsilon_0 \mid \vec{r} \mid} + 2P_p(\vec{r}) \tag{2.5.1}$$

is minimum. Therefore, the following relation is required at $\vec{r} = \pm\vec{r}_c$,.

$$\frac{\partial P_{tot}}{\partial \vec{r}} = 0 \tag{2.5.2}$$

Assuming

$$P_p(x, y, z) = \frac{k_r}{2}r^2 + \frac{k_z}{2}z^2 r^2 = x^2 + y^2, \tag{2.5.3}$$

equation (2.5.2) is obtained by

$$\frac{\partial P_{\text{tot}}}{\partial r} = -\frac{e^2 r}{8\pi\varepsilon_0(r^2 + z^2)^{\frac{3}{2}}} + 2k_r r = 0$$

$$r = 0 \qquad \text{or} \qquad (r^2 + z^2)^{\frac{3}{2}} = \frac{e^2}{16\pi\varepsilon_0 k_r}$$

$$\frac{\partial P_{\text{tot}}}{\partial z} = -\frac{e^2 z}{8\pi\varepsilon_0(r^2 + z^2)^{\frac{3}{2}}} + 2k_r z = 0$$

$$z = 0 \qquad \text{or} \qquad (r^2 + z^2)^{\frac{3}{2}} = \frac{e^2}{16\pi\varepsilon_0 k_z}. \tag{2.5.4}$$

Unless $k_r = k_z$, possible solutions of equation (2.5.4) are

$$\vec{r}_c = (r, z) = (0, d_z) d_z = \left(\frac{e^2}{16\pi\varepsilon_0 k_z}\right)^{\frac{1}{3}}$$

$$P_{\text{tot}}(\vec{r}_c) = P_{z-\text{tot}} = 3\left(\frac{e^2}{16\pi\varepsilon_0}\right)^{\frac{2}{3}} k_z^{\frac{1}{3}} \tag{2.5.5}$$

and

$$\vec{r}_c = (d_r, 0) d_r = \left(\frac{e^2}{16\pi\varepsilon_0 k_r}\right)^{\frac{1}{3}}$$

$$P_{\text{tot}}(\vec{r}_c) = P_{r-\text{tot}} = 3\left(\frac{e^2}{16\pi\varepsilon_0}\right)^{\frac{2}{3}} k_r^{\frac{1}{3}}. \tag{2.5.6}$$

When $k_z < k_r$, $P_{z\text{-tot}}$ is the minimum ion energy. When the ion energy is lower than $P_{r\text{-tot}}$, the position of two ions cannot exchange and a string crystal is formed on the z-axis.

Bluemel *et al* indicated that the motion of two ions is chaotic with the kinetic energy of ions are much higher than $P_{r\text{-tot}}$ [14]. Chaos means that the motion is unpredictable because the nonlinearity of motion equation makes the solution of time expansion sensitive to the slight difference of initial condition. In this case, the ions can get close each other, and the interionic Coulomb interaction is very sensitive to the phase of the micromotion. Once the kinetic energies of ions are reduced and a string crystal is formed, the motion of ions is periodic, because the amplitude of the micromotion is negligibly small in comparison with the interionic distance (order of $2d_z$).

To form a stable string crystal with several ions, the following conditions are required. When the position of an ion is shifted in the r-direction, the neighboring ions with the distance of d_i gives a Coulomb force in the r-direction [15]. To maintain a stable string crystal, the following relation is required (see figure 2.11).

$$k_r r > \frac{e^2}{4\pi\varepsilon_0}\frac{2r}{d_i^3}, \qquad d_i > d_{min} = \left(\frac{e^2}{2\pi\varepsilon_0 k_r}\right)^{\frac{1}{3}}. \tag{2.5.7}$$

For the ion at the position of $z_i(>0)$ the Coulomb force from the inside neighboring ion and the sum of the Coulomb force from outside neighboring ion and the trap force should be balanced as shown in figure 2.12. This relation is given by

$$\frac{e^2}{4\pi\varepsilon_0(z_i - z_{i-1})^2} = \frac{e^2}{4\pi\varepsilon_0(z_{i+1} - z_i)^2} + k_z z_i. \tag{2.5.8}$$

Equation (2.5.8) shows $z_{i+1} - z_i > z_i - z_{i-1}$. The minimum value of $z_{i+1} - z_i$ (around $z = 0$) must be larger than $d_{\min}$. With a certain number of i, equation (2.5.8) cannot be satisfied because of

$$\frac{e^2}{4\pi\varepsilon_0(z_i - z_{i-1})^2} < k_z z_i, \tag{2.5.9}$$

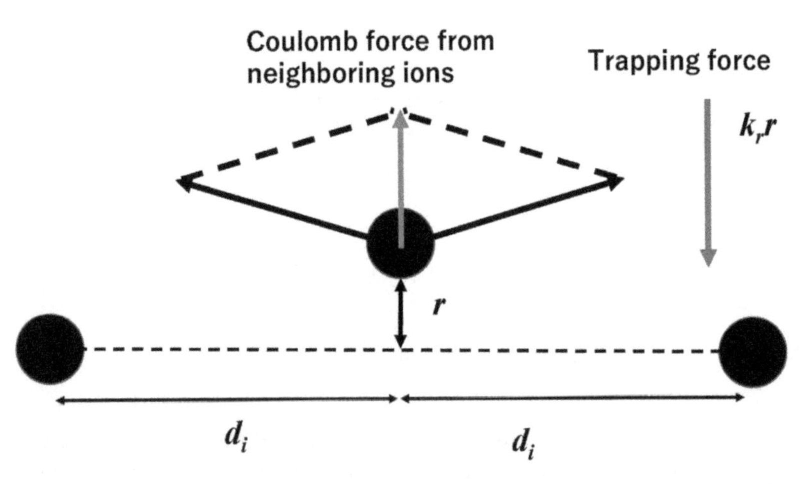

Figure 2.11. To remain a string ion crystal, the minimum interionic distance is required so that the Coulomb force in the radial direction is less than the trap force.

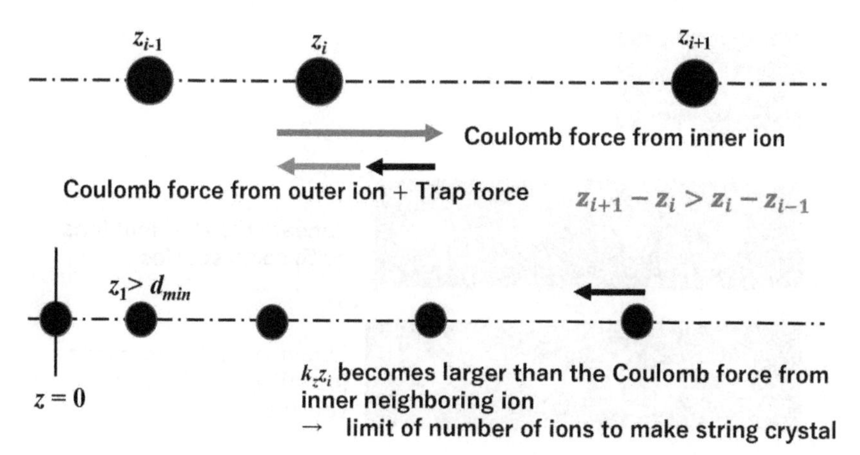

Figure 2.12. The interionic distance becomes larger at the outer side from the balance between Coulomb forces from neighboring ions and trap force. The trap force becomes larger than the Coulomb force from the inner neighboring ion at a position, which makes the limit of number of ions, with which a string crystal can be formed.

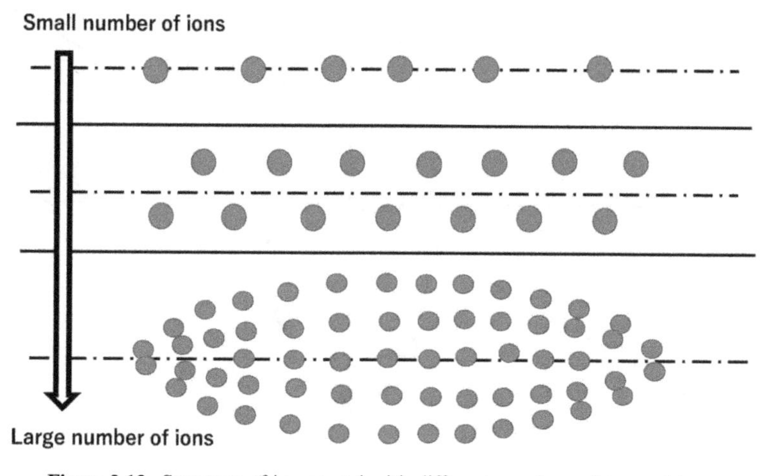

Figure 2.13. Structure of ion crystal with different numbers of trapped ions.

which makes the limit of the number of ions from which a string crystal can be formed.

As the number of trapped ions increases, the structure of the ion crystal changes as shown in figure 2.13.

The fluorescence from a cycle transition (laser induced excitation + spontaneous emission transition) is observed even from a single ion. Figure 2.14 shows the image of a trapped single ion and a linear crystal of four ions. When ions with different species are included in a crystal they are observed as dark spots (no fluorescence) in the crystal.

When ions are trapped in the *x*,*y*-directions using multipole electrodes (figures 1.11 and 1.12), a small number of ions form a ring type crystal as shown in figure 2.15 [16].

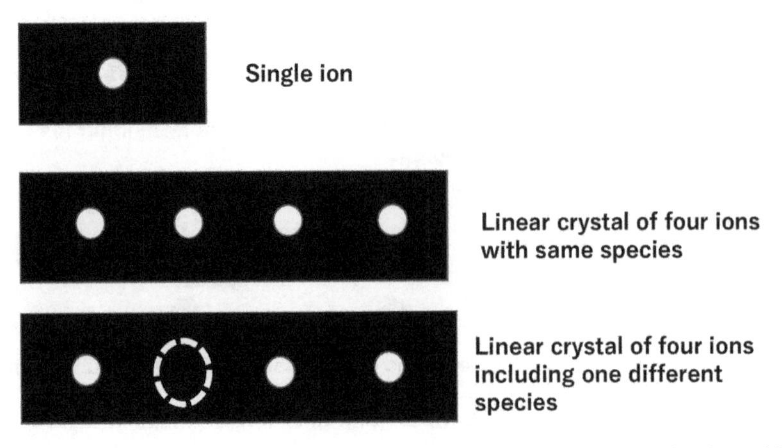

Figure 2.14. Image of ion crystals observed with fluorescence from a cycle transition. When ions with different species are included in the crystal, they are observed as dark spots.

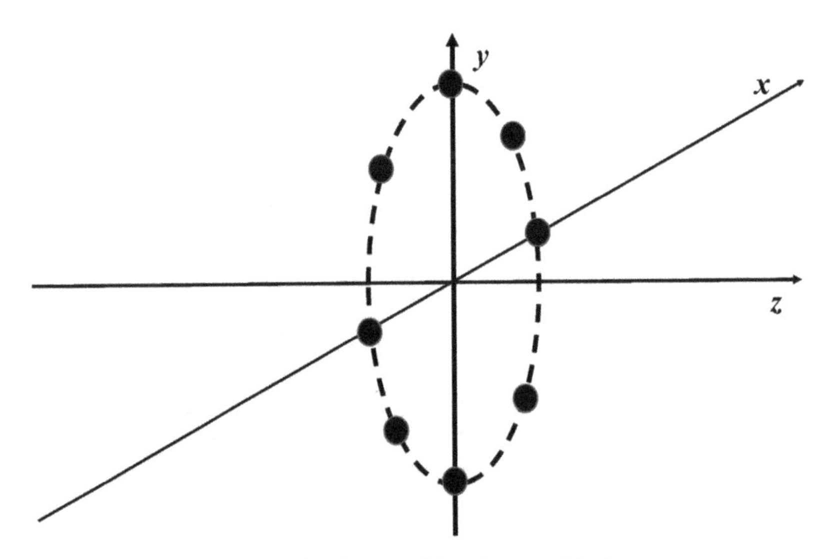

Figure 2.15. Ring ion crystal in a linear multipole trap.

2.6 Trapping of ions by optical dipole force

The tight trapping force makes it possible to trap ions for a long period. On the other hand, trapped ions are confined tightly and the minimum motion energy for one mode (see appendix C) is given by $h\nu_m^\alpha/2$, where ν_m^α is each vibrational motion frequency. There is also the micromotion energy with the RF trap, and it is difficult to attain the vibrational motion energy below 10^{-5} K.

When ions are trapped by the dipole force of a standing wave of laser light (Stark energy shift induced by the electric field of laser light shown in appendix B), much lower kinetic energy is feasible. For example, sympathetic cooling of a single ion to 10^{-7} K with atoms in the Bose–Einstein condensation state (appendix D) was attained [17]. Ions with the kinetic energy below the order of microkelvin are useful for preparation of entangled state between ions (chapter 3) or study of chemical reaction (chapter 4).

Schneider *et al* [18] described the first implementation of an optical dipole trap of ions. Ions are once trapped by the rf-electric field and laser cooled. Then laser light for optical trap is irradiated and the RF trap electric field is ramped down to zero. The when the frequency of the laser light ν is close to the transition frequency of the ion ν_0, dipole trap potential is given by (appendix B)

$$P_{\text{dipole}} = -\frac{h\Omega_R^2}{4\Delta}, \qquad \Delta = \nu_0 - \nu \tag{2.6.1}$$

where Ω_R is the Rabi frequency (defined in equation (A1.5)). When $\nu < \nu_0$, ions are trapped at the anti-node of the standing wave of laser light.

Ions are accelerated by a stary DC electric field, therefore, the voltage applied to the electrode to cancel the electric field (the residual force below 10^{-20} N).

There is limit of period that ions are trapped by the dipole force, because there is a heating effect by the cycle of laser induced excitation and spontaneous emission of photons in random direction (emission of photon gives an acceleration to the ion in the inverse direction). The transition rate is given by

$$R_{\text{tr}} = \frac{\Gamma\Omega_R^2}{\Delta^2}, \tag{2.6.2}$$

where Γ is the spontaneous emission rate. To get high trap potential suppressing the heating effect, it is preferable to take large frequency detuning, because the trap potential is inverse proportional to Δ while the transition rate is inverse proportional to Δ^2.

Schneider *et al* [18] describe the experiment to trap a $^{24}Mg^+$ ion using a laser light with the wavelength of 280 nm. Taking the frequency detuning of 275 GHz, power 190 mW, and the waist radius of 7 μm, the potential depth of 38 mK was obtained. About 80% of ions were trapped with this dipole trap (with the power of 100 mW, 75%). The half-life 2.4 ms of ions in the dipole trap was derived from this experiment. The lifetime in the dipole trap can be improved by using the three-dimensional optical lattice with $\nu > \nu_0$, because ions are trapped at the node of the standing wave and the heating effect is suppressed.

2.7 Optical manipulation and entropy

In this chapter, optical manipulation of trapped ions was introduced. Optical pumping and laser cooling reduce the entropy of trapped ions, getting uniform quantum state and vibrational motion energy. But the increase of entropy is fundamental of thermal dynamics.

The increase of the entropy is obtained by transforming laser light with a uniform phase to a fluorescence light with a random phase. To understand laser cooling, we can interpret the laser light as a cold bath and the scattering of light is interpreted as the heating of photons.

References

[1] Urabe S 1995 *J. IEE Japan* **115** 356 (in Japanese)
[2] Zhang J *et al* 2016 arXiv:1609.03667v1
[3] Rosenband T *et al* 2007 *Phys. Rev. Lett.* **98** 220801
[4] Chou C W *et al* 2010 *Phys. Rev. Lett.* **104** 070802
[5] Ohtsubo N 2017 *Opt. Express* **25** 11725
[6] Born M and Oppenheimer J R 1927 *Ann. Phys.* **389** 457 (in German)
[7] Haensch T W and Shawlow A L 1975 *Opt. Commun.* **13** 68
[8] Wineland D J, Drullinger R E and Walls F L 1978 *Phys. Rev. Lett.* **40** 1639
[9] Raab E L *et al* 1987 *Phys. Rev. Lett.* **59** 2631
[10] Kajita M 2022 *Fundamentals of Analysis in Physics* (Potomac, MD: Bentham Books) 88
[11] Hamann S E *et al* 1998 *Phys. Rev. Lett.* **80** 4149
[12] Roos C F *et al* 2000 *Phys. Rev. Lett.* **85** 554
[13] Germann M *et al* 2014 *Nat. Phys.* **10** 820
[14] Bluemel R *et al* 1989 *Phys. Rev.* A **40** 808
[15] Tachikawa M *et al* 1993 *IEEE Trans. Inst. Meas.* **42** 281
[16] Champenois C *et al* 2010 *Phys. Rev.* A **81** 043410
[17] Zipkes C *et al* 2010 *Nature* **464** 388
[18] Schneider C *et al* 2010 *Nat. Photon* **4** 772

IOP Publishing

Chapter 3

Quantum characteristics of trapped ion

This chapter discusses work into the coupling of different quantum states using trapped ions. Trapped ions are advantageous to manipulate their quantum states over a long period.

From quantum mechanics, the state of a particle (atom, molecule, ion, etc) is given by a wavefunction, having multiple physical values (for example, position, etc) simultaneously. The simultaneous existence of the multiple macroscopic phenomena (called 'Schrödinger's cat state') should be observed using a single particle to confirm this basis of quantum mechanics. A trapped single ion is useful for this purpose.

When two ions A (having states 1 and 2) and B (having states α and β) are co-trapped, coupling state between 'ion A state 1, ion B state α' and 'ion A state 2, ion B state β' can be obtained, this is called the 'entangled state'. With this state, it is possible to get information about one ion by measuring the other ion. The entangled state is realized by the correlation between the ionic internal energy state and the motion energy mode. The motion of both ions cannot be independent because of the Coulomb interaction between them. From the correlation between internal energies of both ions and the motion energy, the internal energy states of both ions can be correlated.

The interaction between a single ion and single photon has also been demonstrated. This phenomenon indicates the entangled states between the quantum energy state of ion and the photon number.

The entangled state between the quantum energy state of ion and its motion mode makes it possible to make a controlled-NOT (CNOT) gate, which is the fundamental of quantum computer.

3.1 Coupling between multi eigenstates

A single trapped ion is useful in order to observe a phenomenon indicating that a particle can have multiple physical values simultaneously, which is a fundamental of quantum mechanics.

doi:10.1088/978-0-7503-5472-1ch3

In quantum mechanics, the properties of all particles are given using wave-function Ψ, and the existence probability is given by $|\Psi|^2$. The physical value X is given by

$$\int \Psi^* \tilde{X} \Psi d\vec{r} \tag{3.1.1}$$

as the average of the given states. Here, $\tilde{X}$ is the operator for X. Wavefunctions with deterministic values of X (called eigenvalue X_e) are called eigenfunctions Ψ_e, and they satisfy the following relation

$$\check{X}\Psi_e = X_e \Psi_e. \tag{3.1.2}$$

All functions can be expressed as a linear combination of eigenfunctions as follows

$$\Psi = \sum a_i \Psi_{ei}. \tag{3.1.3}$$

The following relation is satisfied

$$\int \Psi_{ei}^* \Psi_{ei} d\vec{r} = 1 \int \Psi_{ei}^* \Psi_{ej} d\vec{r} = 0\ (i \neq j)$$

$$\int \Psi^* \tilde{X} \Psi d\vec{r} = \sum |a_i|^2 X_{ei} \sum |a_i|^2 = 1, \tag{3.1.4}$$

and the measurement result of X is one of the eigenvalues X_{ei} with a probability of $|a_i|^2$. Owing to the measurement, the wavefunction changes to the eigenfunction corresponding to the measured eigenvalue. The change in the quantum state by the measurement is called 'quantum destruction'.

Without the measurement of X, the particle has multiple eigenvalues of X simultaneously. For simplicity, we consider the two eigenvalues $X_{0,1}$. The wave-function is given by

$$\Psi = c_0 \Psi_0 + c_1 \Psi_1 \tag{3.1.5}$$

and the probability distribution is given by

$$|\Psi|^2 = |c_0|^2 |\Psi_0|^2 + |c_1|^2 |\Psi_1|^2 + 2Re[c_0^* c_1 \Psi_0^* \Psi_1]. \tag{3.1.6}$$

The third term of the right-hand side of equation (3.1.6) indicates the interference between two states, whose volume integral is zero. The uncertainty principle shows that the product of the uncertainties of the position and the momentum (energy and time) are larger than $h/4\pi$. The position localization (temporal change) of $|\Psi|^2$ is derived from the interference between eigenfunctions with different values of momentum (energy).

The eigenvalues of internal energy of the ion (electron energy, also for molecular ions vibrational rotational motion) are discrete as shown in section 2.1. The eigenvalue of the secular motion energy of a trapped ion is also discrete, which is tresated approximately as harmonic oscillation as shown in appendix C. We can manipulate the coupling between different energy states, for example by the interaction with a laser light (see appendix A). This chapter indicates the phenomena observed with a small number of ions with which the internal energy states are coupled with other physical states. Irradiating a probe laser, the internal energy state of ions can be determined from the information whether the fluorescence signal is observed or not. Section 3.2 shows an example that the different motion modes of the ion are also still coupled after the determination of the internal energy state of a single trapped ion. Section 3.3 gives an example showing that the internal energy states of two ions are correlated. The coupling between the internal energy state of a single trapped ion and photon number in the cavity is shown in section 3.4. These phenomena demonstrate the interpretation of quantum mechanics, based on the coupled states.

3.2 Schrödinger's cat

The combination of multiple physical values can be considered as the phenomena in a microscopic sight because the combined physical values are distributed in a narrow region with the macroscopic sight. For example, the broadening of the atomic wave-packet at room temperature is much less than the atomic size.

However, there is the Schrödinger's cat paradox (figure 3.1). A vessel of a poison gas is placed next to a cat. We can construct a system that the vessel is broken when a radioactive atom decays. When a radioactive atom decays, the cat is killed but without the decay, the cat remains alive. The atom is in the coupled state between 'decayed' and 'not decayed', which is a phenomenon in a microscopic world. But the cat should be in the coupled state between 'dead' and 'alive', although it is a macroscopic phenomenon.

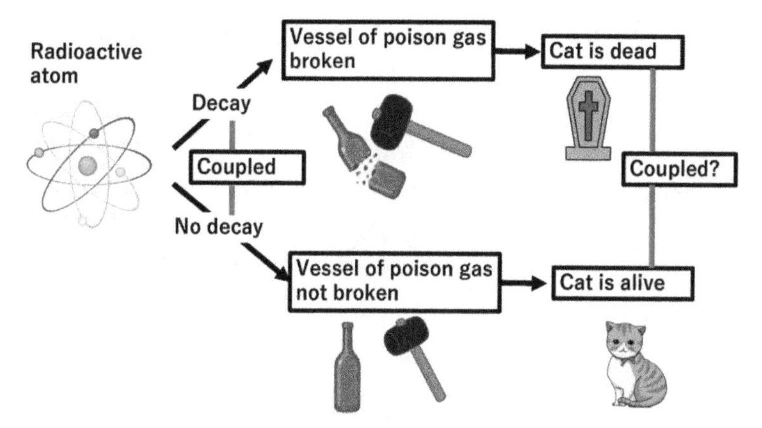

Figure 3.1. The Schrödinger's cat paradox. A vessel of poison gas is put next to a cat. The vessel is broken when a radioactive atom decays, but nothing happens without the decay. The atom is in a coupled state between 'decay' and 'no decay' in a microscopic world. The cat should then be in the coupled state between 'dead' and 'alive', which is a macroscopic phenomenon. This paradox shows that the coupled state is possible also with macroscopic phenomena.

This paradox shows that the coupling between different states should also be observed in a macroscopic world, which is called the Schrödinger's cat phenomenon.

The Schrödinger's cat phenomenon should be observed with a single particle. Observing multiple particles, the coupling between multiple states is observed as the statistic distribution.

Monroe *et al* observed the Schrödinger's cat phenomenon using a single ion, by inducing the vibrational motion with two inverse phases ($x = \pm x_0 \sin[2\pi\nu_v t]$) [1]. The amplitude of the vibrational motion x_0 is much larger than the broadening of the wave-packet, then the wave-packet is localized at two distant places (in maximum $2x_0$) simultaneously. When both parts of the wave-packet overlap at $x \approx 0$, an interference is observed, showing that we cannot distinguish which part of the wave-packet is real.

How can we induce a vibrational motion with two phases? The temporal vibrational motion is not induced by the transition between different eigenstates of motion energy, because the distribution of the wavefunction of eigenstate of vibrational motion cannot have any temporal change (uncertainty principle between the energy and time). The motion shown in figure 3.2 is attained when different eigenstates of motion energy (given by n_{vib}) are coupled. This motion is induced giving a force using two counterpropagating laser lights with the electric field of (see figure 3.3)

$$E_1 = E_0 \exp\left[2\pi\nu_1 i\left(t - \frac{x}{c}\right)\right]$$

$$E_2 = E_0 \exp\left[2\pi\nu_2 i\left(t + \frac{x}{c}\right)\right]$$

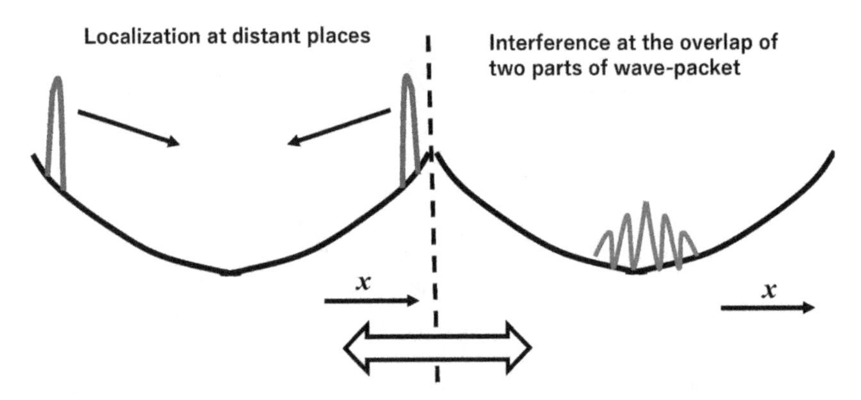

Figure 3.2. Schrödinger's cat phenomenon observed using a single trapped ion [1]. The vibrational motion with two inverse phase is induced, then the wave-packet is localized at two different places with the distance much larger than the broadening of the wave-packet. When both parts of wave-packet overlap, an interference is observed.

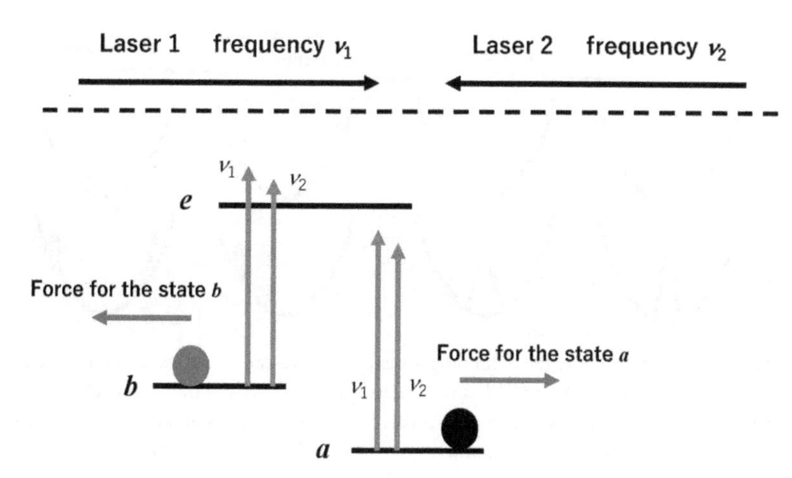

Figure 3.3. The vibration motion is induced by counterpropagating two lasers. The motion direction depends on the quantum energy state.

$$|E_1 + E_2|^2 = 2E_0^2\left(1 + \cos\left[2\pi\left\{(\nu_1 - \nu_2)\left(t - \frac{\nu_1 + \nu_2}{c(\nu_1 - \nu_2)}x\right)\right\}\right]\right). \tag{3.2.1}$$

Here we assume the energy structure with two ground states a and b and one excited state e. The Stark energy shifts in the s (s: a or b) state are given by (see appendix B)

$$\Delta E_S(s) = \frac{2\Omega_{Rs}^2}{\delta\nu_s}\left(1 + \cos\left[2\pi\left\{(\nu_1 - \nu_2)\left(t - \frac{\nu_1 + \nu_2}{c(\nu_1 - \nu_2)}x\right)\right\}\right]\right)$$

$$\delta\nu_s = \nu_{\mathrm{se}} - \nu_1 \approx \nu_{\mathrm{se}} - \nu_2 \quad \Omega_{\mathrm{Rs}} = d_{\mathrm{se}}E_0$$

ν_{se}: s–e transition frequency d_{se}: s–e transition dipole moment

$$|\nu_1 - \nu_2| \ll |\nu_{\mathrm{ae}} - \nu_{\mathrm{be}}| \quad \text{is assumed.} \tag{3.2.2}$$

The temporal change of the velocity of the ion at the position where the trap potential is minimum ($n_{\mathrm{vib}} = 0$) is given by

$$\begin{aligned} \nu(t) &= \int \frac{\partial(\Delta E_S)}{\partial x}\mathrm{d}t \\ &= -\frac{2\Omega_{Rs}^2}{\delta\nu_s}\frac{\nu_1 + \nu_2}{c(\nu_1 - \nu_2)}\cos\left[2\pi\left\{(\nu_1 - \nu_2)\left(t - \frac{\nu_1 + \nu_2}{c(\nu_1 - \nu_2)}x\right)\right\}\right]. \end{aligned} \tag{3.2.3}$$

When $\delta\nu_a > 0$ and $\delta\nu_b < 0$, the ions in the a and b states have vibrational motions in different directions as shown in figure 3.3. We call these different motion modes $\mathrm{Vib}_{1,2}$.

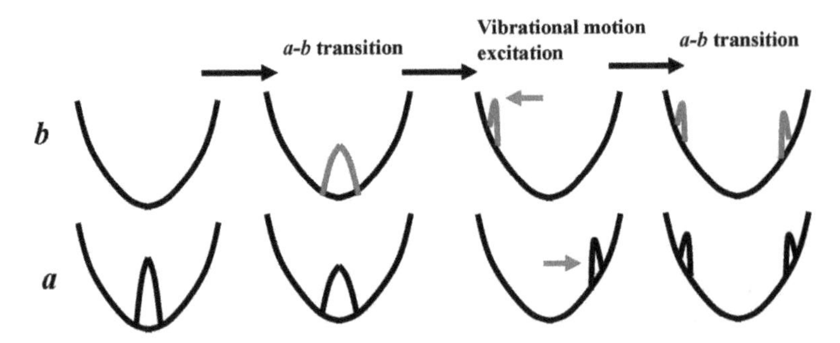

Figure 3.4. The procedure to observe the Schrödinger's cat phenomenon. The state is initially localized to the state a. The coupled state between the a and b states is constructed by the a–b transition. Then the vibrational motion is excited using two laser lights, which gives forces to the a and b states in the inverse directions. Giving the a–b transition again, both vibrational motion modes are coupled in the a and b states.

The procedure to observe the Schrödinger's cat state is shown as follows (see figure 3.4), assuming the initial state to be the $\Phi_0 = (a, n_{\text{vib}} = 0)$ state. By the a–b transition (appendix A),

$$\Phi_1 = \cos\left(\frac{\varphi}{2}\right)(a, n_{\text{vib}} = 0) + \sin\left(\frac{\varphi}{2}\right)(b, n_{\text{vib}} = 0) \tag{3.2.4}$$

When the vibrational motion is excited with this state, the state is transformed to

$$\Phi_2 = \cos\left(\frac{\varphi}{2}\right)(a, \text{Vib}_1) + \sin\left(\frac{\varphi}{2}\right)(b, \text{Vib}_2). \tag{3.2.5}$$

With this state, the wave-packet can be distributed in two places with the distance being much larger than the broadening of the wave-packet. But the Schrödinger's cat state cannot be observed by the fluorescence from the cycling transition from the a or b state, because the internal energy state converges to one of a or b states and the motion mode also converges to one of the Vib_1 or Vib_2 states. The Schrödinger's cat state is observed after the following state is constructed by inducing the a–b transition again.

$$\begin{aligned}\Phi_3 = {} & \cos\left(\frac{\varphi'}{2}\right)\left[\cos\left(\frac{\varphi}{2}\right)(a, \text{Vib}_1) + \sin\left(\frac{\varphi}{2}\right)(a, \text{Vib}_2)\right] \\ & + \sin\left(\frac{\varphi'}{2}\right)\left[\cos\left(\frac{\varphi}{2}\right)(b, \text{Vib}_1) + \sin\left(\frac{\varphi}{2}\right)(b, \text{Vib}_2)\right].\end{aligned} \tag{3.2.6}$$

Observing the fluorescence from the cycling transition, the internal energy state of the ion is determined to be one of the a or b states and the wavefunction converges to

$$\begin{aligned}\Phi'_3 = {} & \cos\left(\frac{\varphi}{2}\right)(a, \text{Vib}_1) + \sin\left(\frac{\varphi}{2}\right)(a, \text{Vib}_2) \text{ or} \\ & \cos\left(\frac{\varphi}{2}\right)(b, \text{Vib}_1) + \sin\left(\frac{\varphi}{2}\right)(b, \text{Vib}_2).\end{aligned} \tag{3.2.7}$$

Two motion modes are coupled with this state, which is indicated as the Schrödinger's cat state. As shown in [1], the Schrödinger's cat state exists, but it is difficult to observe because it decays within a period shorter than 10^{-5} s.

The Schrödinger's cat state is possible, not only with the one-dimensional vibrational motion, but also with the rotational motion in the counterpropagating directions. The overlapping of two parts of a wave-packet is observed as the interference, which is useful to construct a gyroscope (measurement of the rotational angular velocity using the Sagnac effect) with high sensitivity and small size [2].

3.3 Entangled state

The coupling between multiple states is one of the most distinctive properties of quantum mechanics. We now discuss the coupling of the two states of two ions A and B

$$\Phi_A = c_1\phi_1 + c_2\ \phi_2$$

$$\Phi_B = c_\alpha\phi_\alpha + c_\beta\ \phi_\beta. \tag{3.3.1}$$

When there is no correlation between the states of both ions, the total state is given by

$$\Phi_{\text{tot}} = \Phi_A\Phi_B = c_1\ c_\alpha\phi_1\phi_\alpha + c_1\ c_\beta\phi_1\phi_\beta + c_2\ c_\alpha\phi_2\phi_\alpha + c_2\ c_\beta\phi_2\phi_\beta. \tag{3.3.2}$$

However, there is also a state called the 'entangled state'

$$\Phi_{\text{ent}} = C_{1\alpha}\phi_1\phi_\alpha + C_{2\beta}\phi_2\phi_\beta. \tag{3.3.3}$$

If we measure ion A to be the 1 state after making an entangled state, we know that ion B will be in the α state without measuring ion B. Quantum teleportation is a technology to make an entangled state between multi-particles with which we can know information on all particles by measuring one particle [3].

How then can we give the correlation between quantum states of two ions? Most important is that the motion mode is not independent of both ions because of the Coulomb interaction, as shown in appendix C. When the motion of one ion is excited, the motion of the other ion is also excited.

To manipulate the motion energy mode, the kinetic energy of ions should be reduced so that motion energy state in one of the modes is in the lowest eigenstate; for example, the relative motion in the z-direction with the frequency of ν_m^R. The energy eigenvalues are given by $E_{m-\text{vib}}^R = (n_{\text{vib}}^R + 1/2)h\nu_m^R$, where n_m^R is the quantum number to give the motion energy. The entangled state is given with the following procedures for example (see figure 3.5)

(1) as the initial state, $\Phi = \langle\Phi_A|n_{\text{vib}}^R|\Phi_B\rangle = \langle\phi_1|0|\phi_\alpha\rangle$ is prepared
(2) with ion A, the 1→2 blue sideband transition ($\Delta n_{\text{vib}}^R = 1$, see section 2.4.2) is induced: $\Phi = c_1\langle\phi_1|0|\phi_\alpha\rangle + c_2\langle\phi_2|1|\phi_\alpha\rangle$
(3) with ion B, the $\alpha \to \beta$ red sideband transition ($\Delta n_{\text{vib}}^R = -1$, see section 2.4.2) is induced: $\Phi = c_1\langle\phi_1|0|\phi_\alpha\rangle + c_2\left\langle\phi_2|0|\phi_\beta\right\rangle$

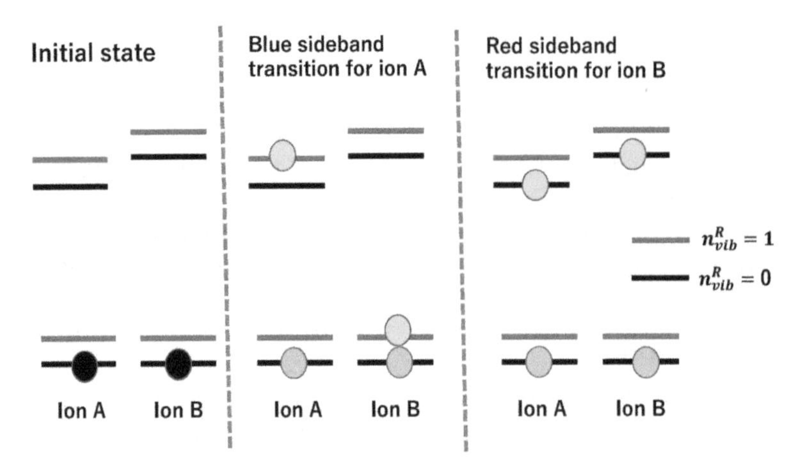

Figure 3.5. One example to produce an entangled state using a blue sideband transition of ion A and red sideband transition of ion B.

For example, the entangled state is used to monitor the quantum state of an ion with which the fluorescence from a cycle transition is not observed.

When this method is used for the precision measurement of the $1 \rightarrow 2$ transition frequency of ion A (chapter 5), the procedure (2) is performed by 'induction of the carrier $1 \rightarrow 2$ transitions' and 'excitation of motion energy by the red sideband $(2, n^R_{\text{vib}} = 0) \rightarrow (1, n^R_{\text{vib}} = 1)$' [4]. The total procedure is given by:

$$\left\langle \phi_1 | 0 | \phi_\alpha \right\rangle \rightarrow c_1\left\langle \phi_1 | 0 | \phi_\alpha \right\rangle + c_2\left\langle \phi_2 | 0 | \phi_\alpha \right\rangle \rightarrow c_1\left\langle \phi_1 | 0 | \phi_\alpha \right\rangle + c_2\left\langle \phi_1 | 1 | \phi_\alpha \right\rangle$$
$$\rightarrow c_1\left\langle \phi_1 | 0 | \phi_\alpha \right\rangle + c_2\left\langle \phi_1 | 0 | \phi_\beta \right\rangle.$$

The state detection of ion B is performed observing the fluorescence from a cycle transition. When ion B is in the β state, the transition of ion A was induced at the first step. After pumping ion B back to the α state, the measurement cycle can be repeated.

When two ions are the same species having two energy states g and e, the state the coupling between the $\langle g, g, n^R_{\text{vib}} = 0\rangle$ and the $\langle e, e, n^R_{\text{vib}} = 0\rangle$ states is obtained by the two-photon transitions using two laser lights with the frequencies of $\nu_0 \pm (\nu^R_{\text{vib}} - \delta_R)$, where ν_0 is the g–e transition frequency. The two-photon transition is induced taking $\langle g, e, n^R_{\text{vib}} = 1\rangle$ or $\langle e, g, \nu^R_{\text{vib}} = 1\rangle$ states as the intermediate state. An arbitral frequency detuning δ_R is given to suppress the one photon transition to the intermediate states. With this method, the entangled state was also obtained with four ions [5, 6].

3.4 Interaction between a single trapped ion and a single photon in a cavity

In a cavity, the number of states of light is quite different from that in a free space. For particles (atoms, molecules, ions) in a cavity, the spontaneous emission rate can be increased or decreased by tuning the cavity length [7]. Black body radiation

(BBR) from a finite size particle has the selectivity of the mode from its size, because it has a role as a cavity [8]. The research field investigating the interaction between a matter and light in a cavity is called cavity quantum electrodynamics (CQED).

It is useful to observe the interaction between a single ion trapped in a cavity and one photon. For an ion with two energy states α and β, the coupling state $\Phi = \cos(\varphi/2)|\alpha, n_p = 1\rangle + \sin(\varphi/2)|\beta, n_p = 0\rangle$, where n_p is the photon number. Keller *et al* generated continuous single photons using an ion-trap cavity system [9]. A single $^{40}Ca^+$ ion is trapped inside an optical cavity (length 8 mm), which is resonant with the $^2D_{3/2}$ (α)–$^2P_{1/2}$ (β) transition. Irradiating a pump laser, resonant to the $^2S_{1/2}$–$^2P_{1/2}$ transition, a cavity assisted Raman transition is induced and a single photon with the wavelength of 866 nm is supplied to the cavity, as shown in figure 3.6. The photon pulse is emitted from one mirror with a transmissivity of 600 ppm. The Rabi frequency of the α–β transition $\Omega_R/2\pi$ (defined in appendix A), damping rate of the photon in the cavity $\kappa/2\pi$, and the spontaneous decay rate $\gamma/2\pi$ were 0.92, 1.2 MHz, and 1.69 MHz, respectively.

$\Omega_R \gg \kappa, \gamma$ is required to improve the efficiency of one photon generation (called strong coupling). The Rabi frequency is given by $\Omega_R = d_{\alpha\beta}E_L$ ($d_{\alpha\beta}$: transition dipole moment, E_L: amplitude of light electric field), and the photon energy is given by $h\nu_L = \varepsilon_0 E_L^2 V_{\text{mode}}/2$ (ν_L: light frequency, V_{mode}: volume of the area of light in the cavity). Therefore, $\Omega_R \propto 1/\sqrt{V_{\text{mode}}}$ is derived. The short length of the cavity is preferable to obtain high Rabi frequencies, but the influence of the electric charge on the mirror surface becomes more significant. Takahashi *et al* constructed a cavity with a length of 0.37 mm, using optical fibers inserted into the end-cap trap electrode, as shown in figure 3.7 [10, 11]. The concavity edge of the optical fiber was made by CO_2 laser ablation and dielectric coating. The Rabi frequency was increased to 12 MHz and strong coupling was attained. Observing the photon from the cavity (optical fiber), the state is determined to be the α ($^2D_{3/2}$) state. On the other hand, the observation of the fluorescence with the same wavelength of the

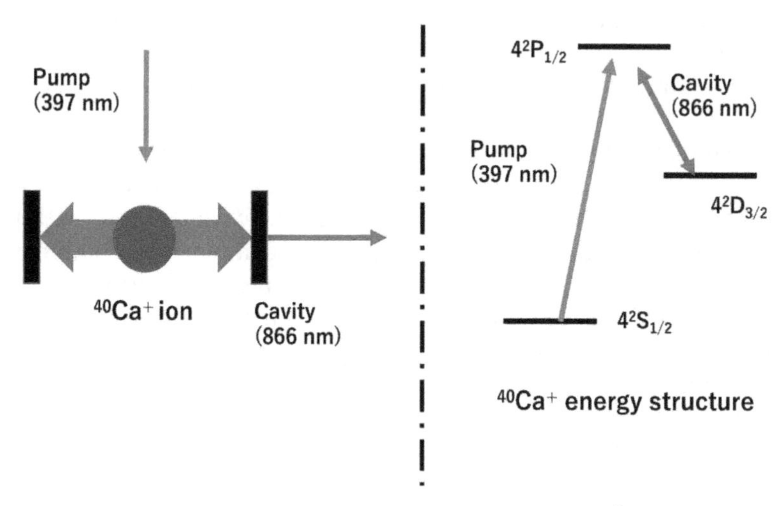

Figure 3.6. Single photon generator using a single trapped $^{40}Ca^+$ ion.

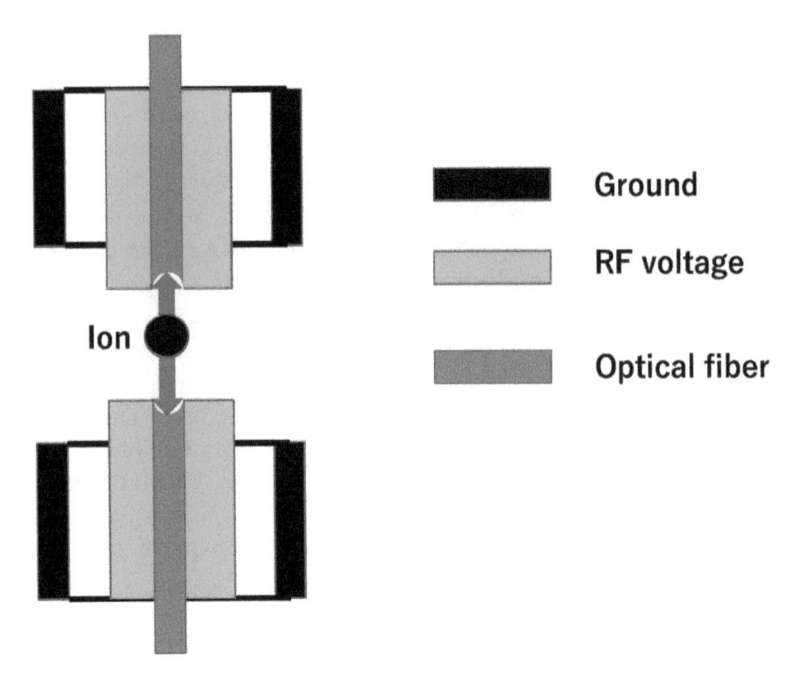

Figure 3.7. The optical cavity formed by optical fibers inserted to the end-car trap electrode.

pump laser (397 nm), the state is determined to the $\beta(^2P_{1/2})$ state. Observing the photon from the cavity and the fluorescence simultaneously, the fluorescence is suppressed when the photon from the cavity is observed. Observing the photon transformed via optical fibers from the cavity, we can also determine the state of the trapped ion at distant places.

3.5 Quantum computer

The fundamental component of a computer is a 'bit', giving two positions '0' or '1'. With the classical computer, these positions are definite. When there are inputs of n_b bits, calculation of 2^{n_b} times (number of combinations of 0 and 1 states for each bit) is required.

With a quantum computer, the bit is called a 'qbit' which can have a coupling of the '0' and '1' states (the simultaneous existence of both states) before the measurement. With the coupled state of '0' and '1' states, the calculation of both positions can be performed simultaneously. Using a quantum computer, all combinations can be calculated with one procedure and the calculation can be performed in a much shorter time and with less electric consumption compared with classical computers (figure 3.8). But there is a serious problem using a coupled state. The coupled state at micro size (atoms or molecules) is destroyed with a slight fluctuation of the circumstance (decoherence). The decoherence was proved experimentally [12], and it creates a barrier to the development of the quantum computer.

The entangled state of trapped ions was expected to be advantageous for the quantum computer because they are isolated from the circumstance [13].

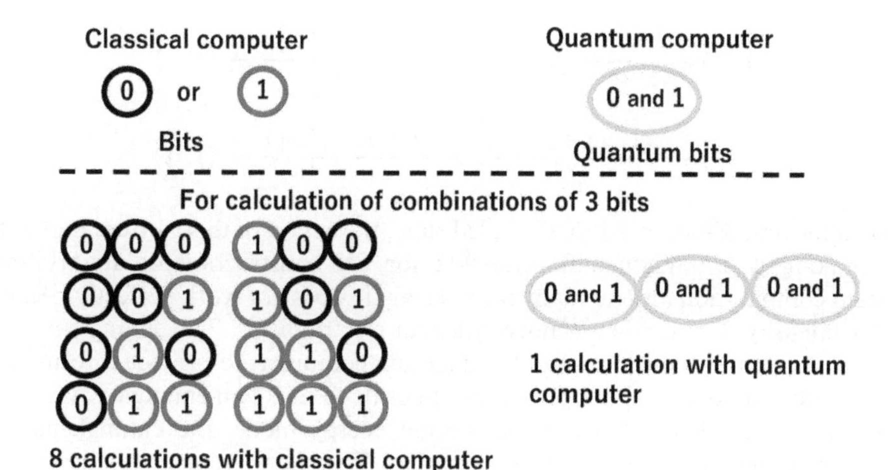

Figure 3.8. The three bits of classical and quantum computers. Eight calculations are required with a classical computer to cover all possible combinations of 0 and 1 with each bit. With a quantum computer, the bit (qbit) can have a coupling of '0' and '1' states, and we can obtaine a result with one calculation. Reproduced from [13].

All computational operations are a combination of the controlled-NOT (CNOT) gates with two q-bits (control bit and target bit) [14]. With a CNOT gate, the following operation is performed with the target bit depending on the control bit.

Control bit 0 Target bit $0\rightarrow0$, $1\rightarrow1$

Control bit 1 Target bit $0\rightarrow1$, $1\rightarrow0$

This gate was realized by Monroe *et al* using a $^9Be^+$ ion, using the motion energy mode n_{vib} (= 0 or 1) as the control bit and the hyperfine energy state of the ion $\downarrow$ $(=|F = 2, m_F = 2\rangle)$ and $\uparrow$ $(=|F = 1, m_F = 1\rangle)$ as the target bit. An additional state aux $(=|F = 2, m_F = 2\rangle)$ is also used to give a phase transition [15]. The CNOT gate is given using the following procedure

(1) $\pi/2$-carrier transition (see appendix A)

(2) 2π-blue sideband transition $\langle n_{\text{vib}}, \uparrow\rangle \rightarrow \langle n_{\text{vib}} - 1, \text{aux}\rangle$ to make the $\langle 1, \uparrow\rangle \rightarrow -\langle 1, \uparrow\rangle$ transition: the transition phase after this procedure $\pi/4$ with $n_{\text{vib}} = 0$ and $5\pi/4$ with $n_{\text{vib}} = 1$

(3) $-\pi/2$-carrier transition: the phase transition $\pi/4\rightarrow0$ for $n_{\text{vib}} = 0$ and $5\pi/4 \rightarrow \pi$ for $n_{\text{vib}} = 1$.

With this procedure,

$$\langle 0, \downarrow\rangle \rightarrow \frac{\langle 0, \downarrow\rangle + \langle 0, \uparrow\rangle}{\sqrt{2}} \rightarrow \frac{\langle 0, \downarrow\rangle + \langle 0, \uparrow\rangle}{\sqrt{2}} \rightarrow \langle 0, \downarrow\rangle$$

$$\langle 0, \uparrow\rangle \rightarrow \frac{\langle 0, \downarrow\rangle + \langle 0, \uparrow\rangle}{\sqrt{2}} \rightarrow \frac{\langle 0, \downarrow\rangle + \langle 0, \uparrow\rangle}{\sqrt{2}} \rightarrow \langle 0, \uparrow\rangle$$

$$\langle 1, \downarrow\rangle \rightarrow \frac{\langle 1, \downarrow\rangle + \langle 1, \uparrow\rangle}{\sqrt{2}} \rightarrow \frac{\langle 1, \downarrow\rangle - \langle 1, \uparrow\rangle}{\sqrt{2}} \rightarrow \langle 1, \uparrow\rangle$$

$$\langle 1, \uparrow\rangle \rightarrow \frac{\langle 1, \downarrow\rangle + \langle 1, \uparrow\rangle}{\sqrt{2}} \rightarrow \frac{\langle 1, \downarrow\rangle - \langle 1, \uparrow\rangle}{\sqrt{2}} \rightarrow \langle 1, \downarrow\rangle$$

When the qbit is initialized to a coupled state, the action of this gate entangles the ions, making it a fundamental operation for the construction of an arbitrary quantum computation among many ions. The gate does not work perfectly because of laser intensity fluctuations or noisy ambient electric fields. The entanglement of trapped ions is important to realize the quantum computer, to provide information about one ion to others. Although it is not realistic to make an entangled state with many ions (more than 20) from the motion energy mode, the entanglement is constructed using photons emitted from ions [16]. The photons can travel through an optical fiber and can make an entangled state between ions at distant places.

References

[1] Monroe C *et al* 1996 *Science* **272** 1131

[2] Shinjo A *et al* 2021 *Phys. Rev. Lett.* **126** 153604

[3] Terada S *et al* 2013 *Nature* **500** 315

[4] Chou C W *et al* 2010 *Phys. Rev. Lett.* **104** 070802

[5] Molmer K and Sorensen A 1999 *Phys. Rev. Lett.* **82** 1835

[6] Sackett C A 2000 *Nature* **404** 256

[7] Bradford M and Shen J-T 2013 *Phys. Rev.* A **83** 063830

[8] Odashima H, Tachikawa M and Takehiro K 2009 *Phys. Rev.* A **80** 041806

[9] Keller M *et al* 2004 *Nature* **431** 1075

[10] Takahashi H *et al* 2013 *New J. Phys.* **15** 053011

[11] Takahashi H *et al* 2020 *Phys. Rev. Lett.* **124** 013602

[12] Haroche S and Raimond J-M 1997 Le chat de Schroedinger prete a l'experience *Recherche* **301** 50

[13] Kajita M 2020 *Cold Atoms and Molecules* (Bristol: IOP Publishing)

[14] Cirac J J and Zoller P 1995 *Phys. Rev. Lett.* **74** 4091

[15] DiVincenzo D P 1995 *Phys. Rev.* A **51** 1015

[16] Monroe C R *et al* 1995 *Phys. Rev. Lett.* **75** 4714

[17] Monroe C R and Wineland D 2008 Quantum computing with ions *Scientific American* (https://scientificamerican.com/article/quantum-computing-with-ions/)

IOP Publishing

Chapter 4

Chemical reaction of trapped ions

In this chapter, the study of the chemical reactions of trapped ions with background gas is introduced. With the previous research, information on chemical reactions was restricted to those of average different states. Using trapped ions, we can observe the slow temporal procedures of chemical reactions of ions in a selected quantum state. The quantum state and the kinetic energy of the collision partner (atoms, molecules) are also manipulated.

4.1 The motivation to study the chemical reaction of trapped ions

The chemical reaction of particles (atoms, molecules, ions, etc) in interstellar space is one of the hottest research subjects for chemists as it could give information about the origin of life. The reactions have been studied using many particles distributed in wide kinetic energy ranges and different quantum states: electronic, vibrational, rotational, and hyperfine states. With the previous research, only information of the average could be obtained. To obtain detailed information of chemical reaction, the reactivity in a selected the quantum state and the kinetic energy should be measured.

In this chapter, research into the chemical reactions of trapped ions is introduced. Using trapped ions, we can observe chemical reactions between trapped ions and background gas (neutral atoms or molecules), which can take several minutes. We can manipulate the quantum energy state of trapped ions, therefore, we can investigate the temporal procedure of chemical reaction of ions in a selected quantum energy state.

Note that the two-body chemical reaction between an atomic ion and an atom $X^+ + Y \rightarrow XY^+$ is not possible because the simultaneous conservation of total energy and momentum components in all directions is not possible. The three-body reactions between one atomic ion and two atoms are possible because the motion of one atom can compensate for the change of the energy and momentum. The two-body reaction between an atomic ion and a molecule is also possible because

doi:10.1088/978-0-7503-5472-1ch4

molecules can be dissociated, and the change of the energy and momentum can be compensated for with the motion of a fragment.

The quantum states and kinetic energy of the buffer gas can be also manipulated, and this opens the study and control chemical processes at an unprecedented level of accuracy.

4.2 Mass spectrum of RF-trapped ions

The mass of trapped ion changes due to the chemical reaction. Therefore, it is important to monitor the masses of trapped ions [1]. Using the Penning trap, the change of mass is monitored as the change of the frequency of the circle motion (section 1.4). The chemical reaction of NH_3^+ molecular ions in a Penning trap and H_2 or D_2 background gas has been studied [2]. However, trapped ions in a Penning trap can be kicked out of the trap area by collisions with background gas. For a study of collisions between trapped ions and buffer gas, an RF trap (section 1.5) is more useful.

The trapped ions with ultra-low kinetic energy are monitored by observing the fluorescence from cycle transitions (laser induced excitation + spontaneous emission transition). When ions undergo chemical reaction, fluorescence is no longer observed. When some ions in a crystal have a chemical reaction, they are observed as dark spots, as shown in figure 2.13.

We can analyse the composition of the ion after the chemical reaction from the information about the mass. Giving a small perturbation of the voltage $V_s \sin(2\pi\nu_m t)$ (ν_m is the secular motion frequency) to one end of the trap electrode, the kinetic energy increases and the fluorescence signal becomes weaker (RF resonance), as shown in figure 4.1. The change in the fluorescence intensity is also observed when ν_m

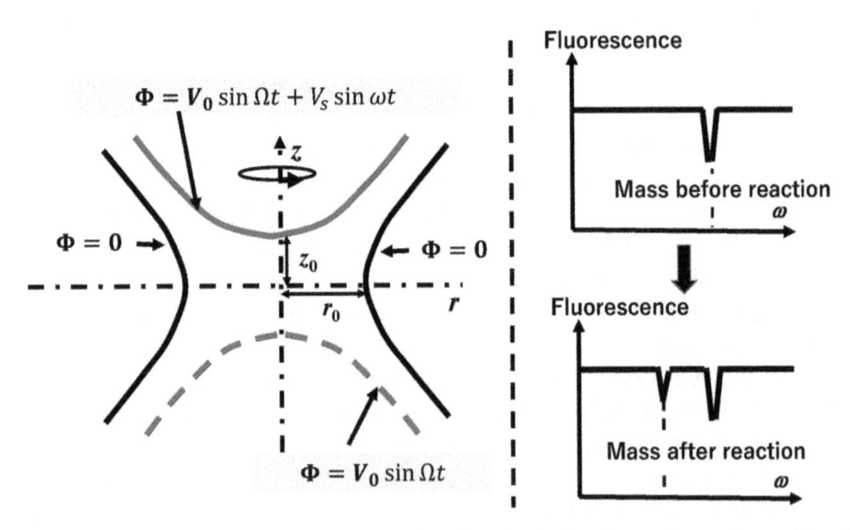

Figure 4.1. Method to monitor the mass of trapped ions by stimulating the secular motion and by giving a perturbation of an additional electric field.

is resonant to the secular motion of the ion which does not give the fluorescence, because the motion excitation of an ion also induces the motion excitation of the surrounding ions by the Coulomb interaction. Equation (1.5.5) indicates that the secular motion frequency is inversely proportional to the mass of the ion, and the mass of the trapped ion is obtained from the frequency of the perturbation voltage where the fluorescence intensity changes. The RF resonance is also observed by the change of the voltage between electrodes (change of the conductance between electrodes). Therefore, mass analysis by RF resonance is possible also when fluorescence is not observed.

4.3 Reaction with H_2 molecules

The QH^+ (Q: ^{24}Mg, ^{40}Ca, ^{88}Sr, ^{138}Ba, ^{174}Yb, ^{202}Hg) and RH^+ (R: ^{4}He, ^{20}Ne, ^{40}Ar) molecular ions have large energy gaps between different vibrational–rotational states (section 2.1) because of the small mass of the H atom. The electric energy structure in the ground state is in the $^1\Sigma$ state (no electron spin and electron orbital angular momentum). Note also that the Q and R nuclear spins are zero. Therefore, these molecular ions have a simple energy structure, and they are useful for the detailed study of molecular energy structures; for example, precision measurement of vibrational–rotational transition frequencies (see chapter 5).

The QH^+ molecular ions are produced from the chemical reaction

$$Q^+ + H_2 \rightarrow QH^+ + H.$$

This reaction is not caused when the Q^+ ion is in the $^2S_{1/2}$ state, because this reaction transition is an endothermic one. When the Q^+ ion is excited to the $^2P_{1/2}$ state (the wavelength of the excitation laser is shown in table 4.1), this reaction can be caused. The reaction rate is higher for the ion with the higher energy of the $^2P_{1/2}$ state; for example, the reaction rate of the Yb^+ ion ($^2S_{1/2}$–$^2P_{1/2}$ transition frequency 369 nm) is higher than the Ca^+ ion ($^2S_{1/2}$–$^2P_{1/2}$ transition frequency 397 nm). This reaction is caused by the procedure of laser cooling because residual H_2 molecules are also included in the background gas without an artificial supply of H_2 gas, and it causes the continuous $^2S_{1/2}$–$^2P_{1/2}$ transition cycle to fade. Sugiyama and Yoda produced $^{174}YbH^+$ molecular ions (confirmed with the mass analysis shown in section 4.2) and dissociated them using laser light with wavelengths of 369.482, 369.202, or 368.947 nm [3]. The lifetime of the ^{138}Ba ion crystal is increased by the photodissociation of $^{138}BaH^+$ molecular ion by a laser with the wavelength of 225 nm [4].

The rare gas ion is in the high energy state and the

$$R^+ + H_2 \rightarrow RH^+ + H$$

reaction is caused without the excitation of the R^+ ion [5]. RH^+ molecular ions are dissociated in the collision with background gas, therefore, the experiment using RH^+ molecular ion should be performed in a cryogenic chamber with the temperature lower than 10 K.

4.4 The reaction between Ca^+ ions and molecules at room temperature

The experiment using alkali-like ions can be disturbed not only by the chemical reaction with H_2 molecules but also by the reaction with other molecules. This chapter demonstrates the chemical reaction rates between $^{40}Ca^+$ ion and H_2O or O_2 molecules.

An experiment was performed to study the reaction rate to the quantum state of $^{40}Ca^+$ ion, when $^{40}Ca^+$ ions are trapped within a H_2O vapor (1.1×10^{-6} Pa) [1]. Figure 4.2 shows the population of $^{40}Ca^+$ ion with irradiating or not irradiating lasers with wavelengths of 397 and 866 nm.

Without laser light irradiation, ions are localized in the $^1S_{1/2}$ state and no fluorescence is observed (figure 4.2(a)). When laser lights with wavelengths of 397 nm and 866 nm, the ions are in the cycling transition and the fluorescence signal is observed with high intensity (figure 4.2(c)). When only the laser light with a wavelength of 397 nm is observed, the population in the $^2D_{3/2}$ state is increased and the fluorescence signal is much weaker because the $^2S_{1/2}$–$^2D_{3/2}$ transition is E1 forbidden (figure 4.2(b)). With procedure (b) there is a $^{40}Ca^+$ ($^2D_{3/2}$) + H_2O chemical reaction. Therefore, the fluorescence intensity with state (c) after procedure (b) is lower than that before this procedure, as shown in figure 4.3. The rate of this chemical reaction was obtained as $(6.7 \pm 0.4) \times 10^{-12}$ cm^3 s^{-1} in the $^2D_{3/2}$ state, and it is negligibly small in the $^2S_{1/2}$ state.

The $^{40}Ca^+ + O_2 \rightarrow {}^{40}CaO^+ + O + \Delta E_c$ reaction rate was measured with different electronic states of $^{40}Ca^+$ ion, where ΔE_c is the change of the energy by the chemical reaction [6]. The $^{40}Ca^+$ is laser cooled to tempuratures of millikelvin and O_2 molecules are at room temperature. Then the average collision energy is of the order of 160 K. Values of ΔE_c at each electronic state of $^{40}Ca^+$ ion is -1.7 eV in the

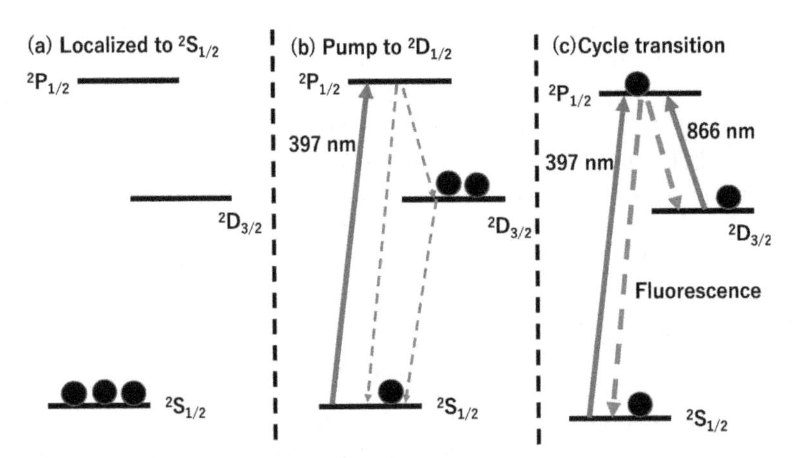

Figure 4.2. Population of $^{40}Ca^+$ ion with three states. (a) Without lasers, the ion is localized in the $^2S_{1/2}$ state. No fluorescence signal is observed. (b) Irradiating only a laser with the wavelength of 397 nm, the population in the $^2D_{3/2}$ state is increased. A weak fluorescence signal is observed. (c) Irradiating two lasers with the wavelength of 397 and 866 nm, a fluorescence signal with high intensity is observed.

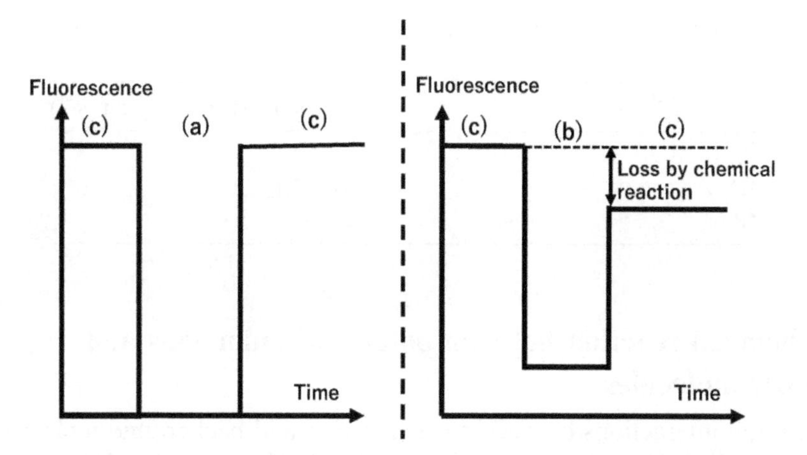

Figure 4.3. Observation of the fluorescence signal from trapped $^{40}Ca^+$ ions. (a) No laser light or ions are localized in $^2S_{1/2}$. (b) Only laser light (397 nm) is irradiated, and the population in $^2D_{3/2}$ is increased. (c) Laser lights (397 and 866 nm) are irradiated to give a intense fluorescence signal. These results shows that the chemical reaction is significant when the population in the 2D3/2 state, because the fluorescence with (c) decreases after having the state (b) for a certain period

Table 4.1. The $^{40}Ca^+ + O_2 \rightarrow {}^{40}CaO^+ + O$ reaction rate with different fractional populations in the $^2P_{1/2}$ and $^2D_{3/2}$ states.

Population in $^2P_{1/2}$	Population in $^2D_{3/2}$	Reaction rate ($\times 10^{-11}$ cm^3 s^{-1})
0.15	0.08	3.3
0.20	0.22	4.6
0.26	0.16	5.1
0.22	0.36	6.0
0.18	0.49	6.1
0.32	0.24	6.3
0.32	0.26	7.6

$^2S_{1/2}$ state, 0.02 eV in the $^2D_{3/2}$ state, and 1.4 eV in the $^2P_{1/2}$ state, respectively. This reaction is caused only when the $^{40}Ca^+$ ion is excited to the $^2P_{1/2}$ and $^2D_{3/2}$ states using lasers with the wavelength of 397 nm and 866 nm, as shown in figure 4.2. The populations in the $^2P_{1/2}$ and $^2D_{3/2}$ states are manipulated irradiating laser lights of 397 nm and 866 nm with different frequencies detuning from the $^2S_{1/2}$–$^2P_{1/2}$ and $^2D_{3/2}$–$^2P_{1/2}$ transitions (Δ_{397} and Δ_{866}). Table 4.1 shows the reaction rate with different fractional populations in the $^2P_{1/2}$ and $^2D_{3/2}$ states.

From the results shown in table 4.1, the reaction rates were obtained as shown in table 4.2.

Table 4.2. The $^{40}Ca^+ + O_2 \rightarrow {}^{40}CaO^+ + O$ reaction rate with different states of $^{40}Ca^+$ ion.

States of $^{40}Ca^+$	Reaction rate ($\times 10^{-10}$ cm^3 s^{-1})
$^2S_{1/2}$	0
$^2D_{3/2}$	0.6 ± 0.1
$^2P_{1/2}$	1.7 ± 0.1

4.5 Chemical reaction between polar molecular ions and polar molecules

The chemical interactions between molecular ions and background molecules have also been studied. Here the $CCl^+ + CH_3CN$ [7], $N_2H^+ + CH_3CN$ [8], and $N_2H^+ + H_2O$ [9] reactions are introduced.

The chemical reaction between CCl^+ molecular ions in a linear trap with the CH_3CN molecules in the background was studied [7]. The CCl^+ molecular ion is relevant in the observation and models of the chemistry of interstellar space. Nitriles are noted for their relevance in prebiotic chemistry, and they have been found in the atmosphere of Saturn's moons and the astmosphere of Titan. CCl^+ molecular ions were produced in a linear trap apparatus by non-resonant multi-photon ionization of C_2Cl_4 molecules in a supersonic molecular beam (seeded in He). The CCl^+ molecular ions produced were sympathetically cooled with co-trapped Ca^+ ion. The chemical reaction was observed by the temporal change of the mass spectrum. The following reactions are caused with the rates of

$CCl^+ + CH_3CN \rightarrow C_2H_3^+ + NCCl + 1.17$ eV reaction rate $(1.6 \pm 0.5) \times 10^{-9}$ cm^3 s^{-1}

$CCl^+ + CH_3CN \rightarrow HNCCl^+ + C_2H_2 + 2.09$ eV reaction rate $(2.2 \pm 0.5) \times 10^{-9}$ cm^3 s^{-1}
which are accompanied by the second reactions

$C_2H_3^+ + CH_3CN \rightarrow CH_3CNH^+ + C_2H_2 + 1.41$ eV reaction rate $(4.2 \pm 1.7) \times 10^{-9}$ cm^3 s^{-1}

$HNCCl^+ + CH_3CN \rightarrow CH_3CNH^+ + NCCl + 0.48$ eV reaction rate $(4.1 \pm 1.2) \times 10^{-9}$ cm^3 s^{-1}

The obtained reaction rates were in good agreement with the average-dipole-orientation (ADO) theory given by Langevin [10].

The $N_2H^+ + CH_3CN \rightarrow CH_3CNH^+ + N_2$ reaction rate was measured with low collision kinetic energy [8]. The N_2H^+ molecular ions are produced by the $N_2^+ + H_2 \rightarrow N_2H^+ + H$ reaction and sympathetically cooled with the co-trapped and laser cooled $^{40}Ca^+$ ions.

CH_3CN molecules are guided by a linear electrode, having the minimum electric field on the axis (figure 4.4). When the molecular beam guide is bent, fast molecules are removed from the waveguide because the centrifugal force overcomes the trapping force. Only slow molecules are guided to the end of the waveguide. With this method, we can select molecules with low kinetic energy. This filtering apparatus also works for

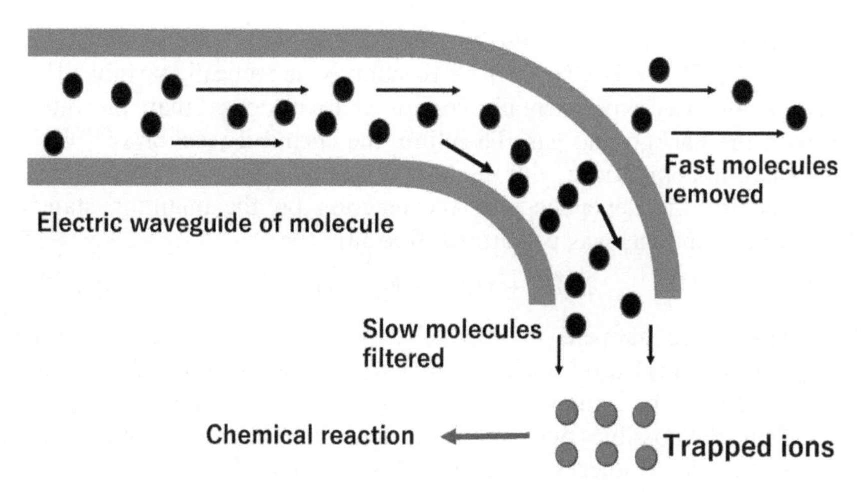

Figure 4.4. The apparatus to guide polar molecules to cause collisions with trapped ions. Only slow polar molecules are filtered using a bent waveguide.

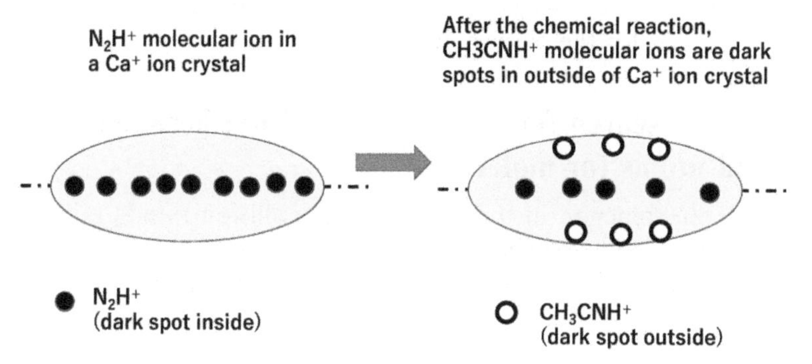

Figure 4.5. The distribution of N_2H^+ and CH_3CNH^+ molecular ions as the dark spots in a $^{40}Ca^+$ ion crystal.

selecting molecules in a single rotational state as the trapping force is strongest for the lowest low field seeking rotational state (positive quadratic Stark energy shift). The temperature of the molecular beam nozzle was cooled to a temperature slightly higher than freezing point so that the fraction of filtered molecules is increased. With the experiment shown in [8], the kinetic energy of the filtered CH_3CN molecules is 3 K.

Both N_2H^+ and CH_3CNH^+ molecular ions are dark spots in a $^{40}Ca^+$ ion crystal. How can we observe the chemical reaction? Equation (1.5.6) shows that the pseudo trap force is inversely proportional to the mass of ions. N_2H^+ ions (smaller mass than $^{40}Ca^+$) are localized inside the $^{40}Ca^+$ ion crystal because of the stronger trap force, while CH_3CNH^+ molecular ions (larger mass than $^{40}Ca^+$) are localized at the outside, as shown in figure 4.5. The rate of the chemical reaction with the collision kinetic energy was estimated to $(1.7\pm0.6) \times 10^{-8}$ cm^3 s^{-1}. Assuming that the reaction rate is proportional to the square root of the collision kinetic energy [10] and the experimental value of the reaction rate with the room temperature, this value is estimated to be 4.0×10^{-8} cm^3 s^{-1}. This discrepancy is an interesting issue to be investigated in future.

Note that CH_3CN molecules can also react with $^{40}Ca^+$ ions, because the $^{40}Ca^+$ ($^2P_{1/2}$) + $CH_3CN \rightarrow$ $^{40}CaH^+$ + CH_2CN + 1.1 eV reaction is energetically possible. However, this reaction rate was experimentally confirmed to be lower than the rate of the reaction with the background gas. Therefore, the chemical reaction of $^{40}Ca^+$ ion is negligibly small in comparison with the N_2H^+ molecular ion.

The search for the dependence of the reaction on the quantum state of the colliding background gas was performed. Recently the

$$N_2H^+ + H_2O \rightarrow N_2 + H_3O^+$$

reaction (which also happens in interstellar space) was studied separating two nuclear spin isomers (H total nuclear spin 1 or 0; called ortho and para) of water molecules [9]. From the symmetry of water molecules exchanging the positions of both H nuclear, the possible rotational states depend on the nuclear spin and there is a difference of the intermolecular potential. A supersonic H_2O beam passes through an electrode with a large electric field gradient. The ortho- and para-molecules obtain different forces from the electric field gradient, and the beams of both isomers are specially separated. Measuring the reaction rate in different places, the reaction rates were obtained to be $(4.0\pm0.9) \times 10^{-9}$ cm^3 s^{-1} for the ortho-molecules and $(5\pm1) \times 10^{-9}$ cm^3 s^{-1} for the para-molecules, respectively.

4.6 Prospect to search for the rate of collision between ultra-cold ions and atoms (or molecules)

The resonance phenomenon in the collision (the collision rate is raised when the collisional kinetic energy matches to the energy gaps of a matter, as shown in figure 4.6) has been theoretically demonstrated. This phenomenon has been observed with neutral cold molecules [11, 12], but it is a new research field to observe it with the collision of ions. Eberle *et al* [13] discussed the procedure to observe the chemical reaction of trapped $^{40}Ca^+$ ions with ultra-cold Rb atoms (laser

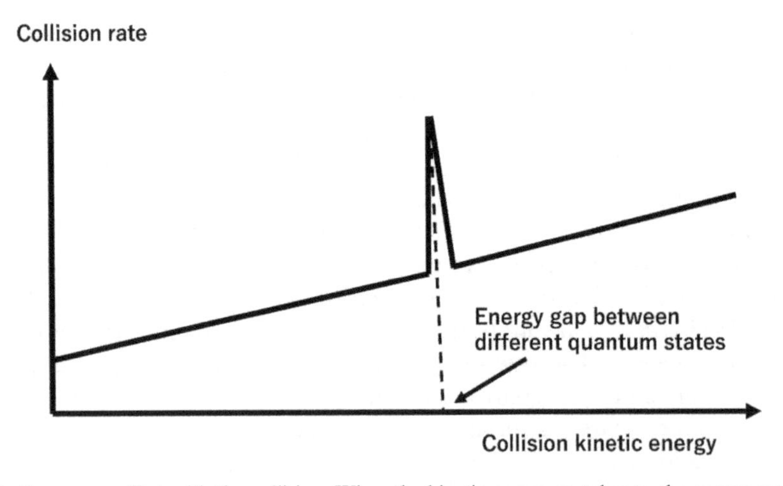

Figure 4.6. Resonance effect with the collision. When the kinetic energy matches to the energy gap between different quantum states, the collision rate is raised.

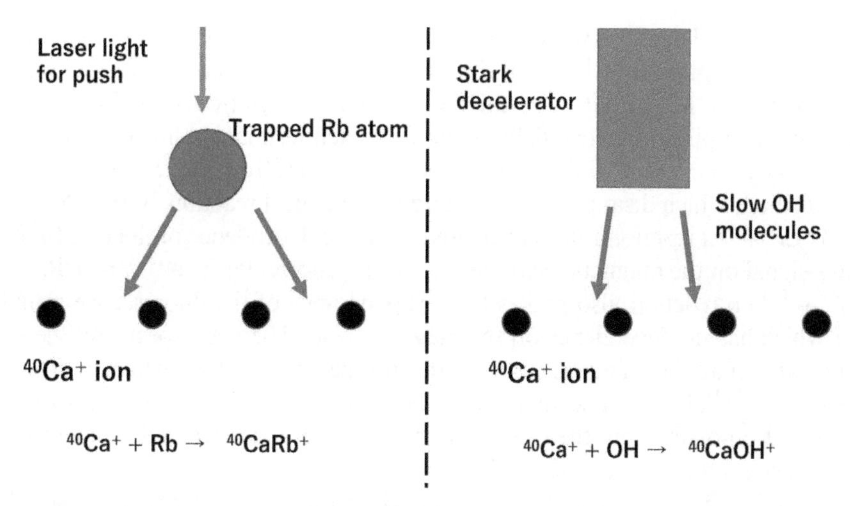

Figure 4.7. Collision between trapped $^{40}Ca^{+}$ ion and cold Rb atoms or OH molecules.

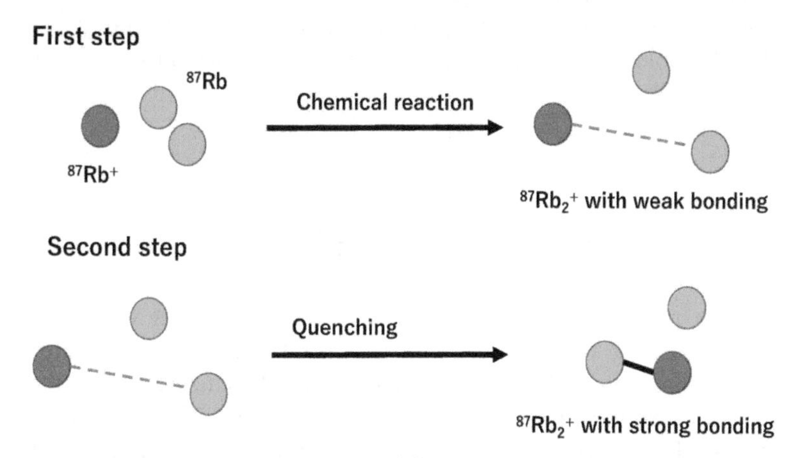

Figure 4.8. The chemical interaction between $^{174}Yb^{+}$ ions and ultra-cold cold $^{6}Li_2$ molecules.

cooled), or OH molecules (passed through a Stark decelerator), as shown in figure 4.7. The Stark molecular decelerator works only for molecules in the lowest low field seeing state, therefore, we can observe the collision with OH molecules in a single quantum state. The collision kinetic energy can be manipulated between 20 and 120 mK with a resolution of a few mK.

Dieterle *et al* [14] showed the reaction between a single $^{87}Rb^{+}$ ion with the Bose–Einstein condensate (BEC, see appendix D) of 8×10^{5} ^{87}Rb atoms. In this experiment, $^{87}Rb^{+}$ ion was produced by the excitation to the Rydberg state and applying a DC electric field. With the three-body interaction $^{87}Rb^{+} + {}^{87}Rb + {}^{87}Rb \rightarrow {}^{87}Rb_2^{+} + {}^{87}Rb$, $^{87}Rb_2^{+}$ a molecular ion with a weak bonding (high vibrational–rotational state) is produced. After that, the two-body molecular ion–atom collision quenches the vibrational–rotational state of the molecular ion towards deeper binding energies, as shown in figure 4.8. This experiment provides important information for the study of the dynamics of ionic impurities in an ultra-cold quantum gas.

Hirzler *et al* [15] demonstrated the $^{174}Yb^+ + {}^6Li_2 \rightarrow {}^{174}Yb^6Li^+ + {}^6Li$ reaction. The $^{174}Yb^+$ ion is trapped in a bath of ^6Li atoms and $^6Li_2$ molecules. A DC magnetic field is applied to the ^6Li bath and the density ratio of ^6Li atoms and $^6Li_2$ molecules depend on the applied magnetic field by the Feshbach resonance shown in appendix E. The existence of the $^{174}Yb^+$ is detected by the fluorescence from the $^2S_{1/2} \leftrightarrow {}^2P_{1/2} \leftrightarrow {}^2D_{3/2}$ cycle transition, which disappears when there is a chemical reaction. The $^{174}Yb^+ + {}^6Li_2 \rightarrow {}^{174}Yb^6Li^* + {}^6Li$ reaction rate is measured from the dependence of the rate to observe the dark signal on the magnetic field (density of $^6Li_2$ molecule). The $^{174}Yb^+ ({}^2P_{1/2}) + {}^6Li \rightarrow {}^6Li^+ + {}^{174}Yb$ reaction also makes the disappearance of the fluorescence signal, the rate of which has no dependence on the magnetic field. The mass spectrum (see section 4.2) indicates that the mass number of the trapped ion is 180, which indicates the production of $^{174}Yb^6Li^*$ molecular ions. The $^{174}Yb^+ + {}^6Li + {}^6Li \rightarrow {}^{174}Yb^6Li^* + {}^6Li$ reaction is also possible, but the rate of this reaction has a quadratic dependence on the atomic density, and is negligible small.

The $^{40}Ca^+ + {}^6Li \rightarrow {}^{40}Ca + {}^6Li^+$ charge exchange collision rate was measured with different collision kinetic energy E_{col} [16]. Langevin's law [10] indicates the formula of the collisional cross section considering the area that the potential energy given by the charge-induced dipole moment interaction is larger than the collisional kinetic energy

$$\sigma_{col} = \pi r_{max}^2 \frac{\alpha_{Li} e^2}{8\pi\varepsilon_0 r_{max}^4} = E_{col}$$

$$\sigma_{col} = \sqrt{\frac{\pi\alpha_{Li} e^2}{2\varepsilon_0 E_{col}}} = \frac{3.3 \times 10^{-28}}{\sqrt{E_{col}(J)}} (m^2), \tag{4.6.1}$$

where α_{Li} is the polarizability of ^6Li atom. The measurement was performed by switching E_{col} between 10 mK and 1 K (change the micromotion amplitude by shifting the trap position adding a DC electric field). The measurement result of the collisional cross sections σ_{col-Ex} were measured with different distributions of the $^2S_{1/2}$, $^2D_{3/2}$, $^2D_{5/2}$, and $^2P_{1/2}$ states of $^{40}Ca^+$ ion (different power and frequency detuning of cooling and repump lasers). The $\sigma_{col-Ex} \propto 1/\sqrt{E_{col}}$ relation was confirmed, but the values were smaller than the estimation by equation (4.6.1). The values of $\sigma_{col-Ex}/\sigma_{col}$ were measured to be 0.15, 0.53, and 0.64 with the $^2D_{3/2}$, $^2D_{5/2}$, and $^2P_{1/2}$ states of $^{40}Ca^+$ ion, respectively. This value was below 10^{-3} with the $^2S_{1/2}$ state. The state dependence of the change–exchange collision rate provides useful information to estimate the potential energy curve between $^{40}Ca^+$ ion and ^6Li atom.

References

[1] Okada K *et al* 2003 *J. Phys. B: At. Mol. Opt. Phys.* **36** 33

[2] Barlow S E and Dunn G H 1987 *Int. J. Mass Spectrom. Ion Processes* **80** 227

[3] Sugiyama K and Yoda J 1997 *Phys. Rev.* A **55** R10

[4] Wu H *et al* 2021 *Phys. Rev.* A **104** 063103

[5] Roth B *et al* 2006 *J. Phys. B: At. Mol. Opt. Phys.* **39** 1241

[6] Schmid P C *et al* 2019 *Mol. Phys.* **117** 3036

[7] Krohn O A *et al* 2021 *J. Chem. Phys.* **154** 074305

[8] Okada K *et al* 2013 *Phys. Rev.* A **87** 043427
[9] Kilaj A *et al* 2018 *Nat. Commun* **9** 2096
[10] Langevin M P 1905 *Ann. Chem. Phys.* **5** 245
[11] Henson A B *et al* 2012 *Science* **338** 234
[12] Chefdeville S *et al* 2013 *Science* **341** 1094
[13] Eberle P *et al* 2015 *J. Phys. Conf. Ser.* **635** 012012
[14] Dieterle T *et al* 2020 *Phys. Rev.* A **102** 041301
[15] Hirzler H *et al* 2022 *Phys. Rev. Lett.* **128** 103401
[16] Saito R *et al* 2017 *Phys. Rev.* A **95** 032709

IOP Publishing

Ion Traps
A gentle introduction
Masatoshi Kajita

Chapter 5

Atomic clocks using trapped ions

Since the development of atomic clocks, time and frequency has been the physical value which can be measured with the lowest uncertainty. This chapter introduces atomic clocks based on the transition frequencies of RF trapped ions. RF trapped ions are localized in a spatial area under the continuous irradiation of probe laser light (or microwave) and it is advantageous to observe the spectrums with narrow linewidth. The measurement uncertainty of the order 10^{-18} has been obtained with $^{27}Al^+$ and $^{171}Yb^+$ transition frequencies. Lower measurement uncertainty might be attained with the transition frequency of highly charged ions. The measurement uncertainty of 10^{-18} might be obtained with vibrational transition frequencies of homonuclear diatomic molecular ions. The rotational transition frequency of heteronuclear molecular ions can be the frequency standard with an uncertainty of 10^{-15} in the THz region. The development of new physics based on precision frequency measurement is also discussed.

5.1 What is an atomic clock?

Physics involves the study of the laws of nature, from which we can make predictions regarding future phenomena. They are established based on measurement results, which have always some uncertainties. New physical phenomena have been discovered when the measurement uncertainties were reduced. Note that the history of the development of physics is well correlated particularly with the developments of new clocks. While the time was measured with solar or water clocks, there was an uncertainty of one hour per day. With this measurement uncertainty of clocks, there was no discrepancy with the Ptolemaic theory to represent the movement of the Sun, Moon, and stars. As the mechanical clock was developed and the measurement uncertainty was reduced to ten minutes per day, some discrepancies were discovered with the positions of stars at a given time. The Copernican theory was established with this background. Newtonian mechanics was established within 100 years since the discovery, by Galielei, of the periodicity of the pendulum's swing and the

doi:10.1088/978-0-7503-5472-1ch5

accuracy of clocks was improved rapidly. The improvement of the accuracy of clocks made it possible to discover the fluctuation of the period taken by Io, one of Jupiter's satellites, to complete an orbit. This was the first phenomenon that indicated the speed of light being a finite value. After the speed of light was measured with a low uncertainty, the identity of the light was clarified to be an electromagnetic wave. The theory of relativity is based on the constancy of the speed of light.

The measurement uncertainty of time and frequency was drastically reduced by the invention of the atomic clock. An atomic clock is a clock based on the transition frequency of atoms or molecules (neutral or ion) [1]. The uncertainty of time and frequency measured using an atomic clock is low enough to detect the relativistic effects of time going in a moving frame or in a gravitational potential. Discoveries of slight effects, which have never been detected with previous measurement uncertainties, are required for the further developments of new physics. Atomic clocks have important roles for this purpose.

Why are atomic clocks so advantageous in comparison with previous clocks; for example, a crystal clock based on the oscillation frequency of a crystal? Simply put, the frequency source of the atomic clock is gaseous phase, while the previous clocks are based on the periodic phenomena of solid materials. The two solid materials cannot have exactly the same structures, because they always contain some impurities. Therefore, two clocks cannot give exactly the same times. The time of each clock also fluctuates because bonding between atoms or molecules in a solid depends on the circumstance (for example, temperature). On the other hand, gaseous atoms and molecules are isolated from each other. The transition frequencies of gaseous atoms and molecules are discrete, and they are determined by the Coulomb force between electron and nucleus, which is a common force for all atoms and molecules. Therefore, the ultra-low measurement uncertainties are obtained with atomic clocks.

This chapter introduces atomic clocks using trapped ions.

5.2 Measurement uncertainty

Measurement uncertainty [2] of time and frequency also exists using atomic clocks. The uncertainties are discussed with 'statistical uncertainty' and the 'systematic uncertainty'. The statistic uncertainty exists because the spectrum has a finite linewidth and the measurement results are not always exact but distribute in the frequency region given by the linewidth, as shown in figure 5.1. The systematic uncertainty exists because the measurement result depends on the circumstance. The real value is defined with a certain circumstance and measurement result with another circumstance shifted from the real value, as shown in figure 5.1. The detailed explanations are given below.

5.2.1 Statistical uncertainty

When we measure something, there is a question of whether it is a real value. Reliability can be confirmed by repeating many times. We will see that the

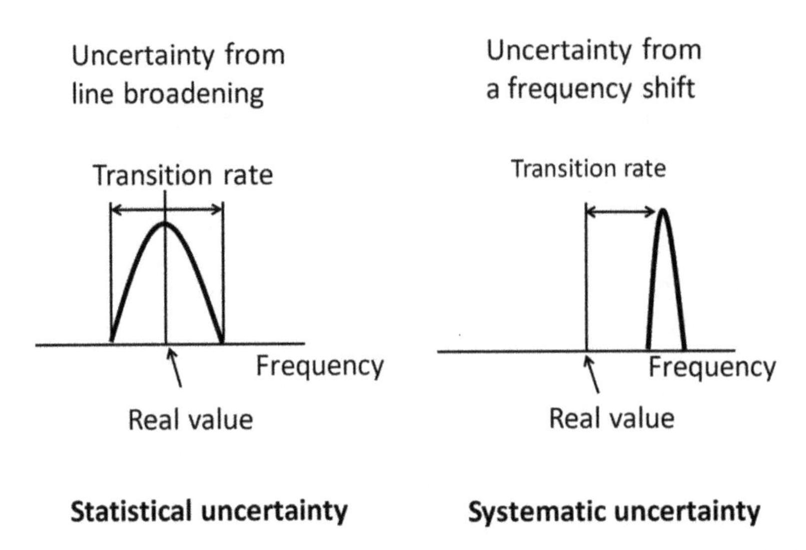

Figure 5.1. Illustration of the statistical and systematic uncertainties. Reproduced from [1].

measurements of the transition frequency ν distribute within a limited frequency area around a central frequency ν_c. The width of the frequency area where the measurement results distribute (spectrum linewidth) is given in the following.

The transition frequency is measured using the phase estimation procedure for ϕ in a finite measurement time τ, where $\phi = 2\pi\nu\tau$. The measurement time is limited by the interaction time between the particle (atom, molecule, ion) and the electromagnetic wave. It can be limited also by a phase jump, by an interatomic collision or in a spontaneous emission transition. The spontaneous emission rate is proportional to the cube of the transition frequency for E1 transition. Therefore, τ is generally limited by the spontaneous emission rate for E1 allowed transitions in the optical region (linewidth of several MHz). For the microwave transitions or E1 forbidden transitions, τ is given by the period of continuous interaction between particles and electromagnetic waves. Assuming a random distribution of the phase at the phase jump,

$$2\pi\nu_c\tau - \frac{1}{2} < \phi < 2\pi\nu_c\tau + \frac{1}{2}$$

$$\nu_c - \Delta\nu < \nu < \nu_c + \Delta\nu, \qquad \Delta\nu = \frac{1}{4\pi\tau}, \tag{5.2.1}$$

where $\Delta\nu$ denotes the spectrum linewidth. This linewidth obeys the uncertainty principle between energy and time, which is one of the most important principles in quantum mechanics.

The statistic uncertainty is reduced by repeating the measurement and taking the average ν_{ave}. The probability distribution of ν_{ave} with the number of measurement samples N_s is given by [3]

$$P(\nu_{\rm ave}) = \frac{\sqrt{N_s}}{(\Delta\nu)\sqrt{\pi}}\exp\left[-N_s\left(\frac{\nu_{\rm ave} - \nu_c}{\Delta\nu}\right)^2\right]. \tag{5.2.2}$$

The statistical measurement uncertainty with the sample number N_S is given by

$$\Delta\nu_{\rm static} = \Delta\nu/\sqrt{N_S}. \tag{5.2.3}$$

The sample number is given by $N_A(\tau_e/T_a)$, where N_A is the number of matters (atoms or molecules) present during the measurement, $\tau_e$the period for a single measurement cycle, and T_a the measurement time. Therefore,

$$\Delta\nu_{\rm statistic} = \Delta f\sqrt{\tau_e/N_A T_a}. \tag{5.2.4}$$

The statistical uncertainty is reduced by measuring many particles over long measurement times. Note also $\tau_e > 1/4\pi(\Delta\nu)$. When $\tau_e \approx 1/4\pi(\Delta\nu)$, the statistical uncertainty is proportional to $\sqrt{\Delta\nu}$.

5.2.2 Systematic uncertainty

We cannot guarantee that the real transition frequency is ν_c. All measurement values can be changed by the circumstance, and the real values should be defined with a certain condition. The transition frequencies are shifted by the electric field (Stark shift), the electric field gradient (electric quadrupole shift), and the magnetic field (Zeeman shift), as shown in appendix B. The theory of relativity indicates that the measurement of frequency ν becomes lower when the matter moves (quadratic Doppler shift). The quadratic Doppler shift is given by

$$\Delta\nu_{\rm QD} = -\frac{K_{\rm E}}{mc^2}, \tag{5.2.5}$$

where $K_{\rm E}$ is the kinetic energy, m is the mass and c is the speed of light. The frequency is shifted by measuring at a place with a different gravitational potential (gravitational red shift). The gravitational red shift is given by

$$\Delta\nu_{\rm GR} = \frac{P_{\rm G}}{mc^2}, \tag{5.2.6}$$

where $P_{\rm G}$ is the gravitational potential energy. The defined transition frequency is the measurement 'with zero electric and magnetic fields, zero velocity, and on the geoid surface'. It is not realistic to measure with the defined condition, therefore, the defined frequency is estimated by correcting the frequency shifts induced by these effects. The uncertainties of the frequency shifts are 'systematic uncertainty', shown in figure 5.1.

5.2.3 Concepts of accuracy and stability

The property of the atomic clock is discussed with the concepts of accuracy and stability. Figure 5.2 shows the results of distributions of measurements with high and low levels of 'accuracy' and 'stability'. Accuracy indicates the reliability of the result after averaging many experimental results and giving corrections to the

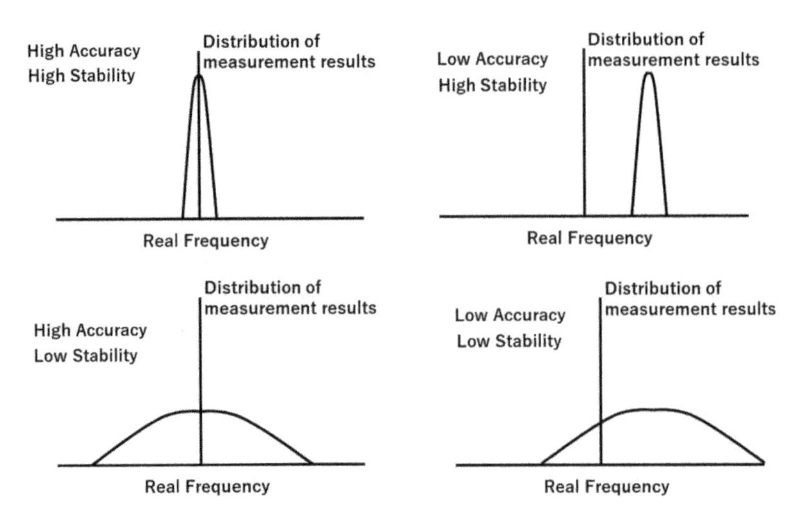

Figure 5.2. Schematic of results for the distribution of measurements with high and low 'accuracy' and 'stability'. Reproduced from [2]. © IOP Publishing Ltd. All rights reserved.

measurement shifts. The accuracy can also be high if the difference between each measurement result is large. To achieve high accuracy, the systematic uncertainty should be small, as the statistical uncertainty can be reduced by repeating the measurement many times. The accuracy is estimated from the possible shift of measurements by the theoretical calculation or experimental measurement with different circumstance. Comparison between different atomic clocks is also performed.

Stability refers to the constancy of each measurement result. The stability can also be high if the measurement results are equally shifted from the real value. While repeating the measurement over a long period of time, there may be temporal changes of the measurement values induced by changes in circumstance. Therefore, we distinguish 'short-term stability' and 'long-term stability' from the averaging time. The former is determined from the statistical uncertainty covering short measurement times, for which the change in circumstance is negligible small. To achieve high 'short-term stability' with the transition frequency, a narrow spectrum linewidth and a large numbers of particles are required. Taking long measurement times, the statistical uncertainty is suppressed, but the influence of the change in conditions (e.g., electric field and magnetic field) becomes significant. Therefore, 'long-term stability' is determined from the systematic uncertainty.

5.3 Special characteristics of atomic clocks using trapped ions

Ions trapped by a Penning trap system cannot be used for atomic clocks, because a strong magnetic field is applied to trapped ions and the Zeeman shift in the transition frequency is significant. On the other hand, ions trapped by a RF electric field are useful for the atomic clock. In a three-dimensional RF trap, a single ion has a secular motion around the position where the electric field is zero. When the amplitude of the secular motion is reduced by laser cooling, this ion is almost free of the trap electric field and the Stark shift (appendix B) in the transition frequency is negligibly small. With the combination of the two-dimensional RF trap and one-dimensional DC trap (linear trap), the RF electric field is zero on the trap axis. When laser cooling is performed with a small number of ions in a linear trap, a string crystal is formed on the trap axis. Measuring the transition frequencies of ions in a string crystal, the Stark shift induced by the trap electric field is negligibly small because the Coulomb force from other ions and the DC trap are balanced. On the other hand, a large electric field gradient is required for the tight confinement. Except for the transition between states without the electric quadrupole moment, the elimination of the electric quadrupole shift (appendix B) is required for precision measurement, which is shown in section 5.5.1.

Recently, precision measurements were made of transition frequencies of neutral atoms trapped by a standing wave of a light (optical lattice clock) [4, 5]. Measurement with more than 1000 atoms is possible with a lattice clock, but the potential depth is of the order of 10^{-5} K and the trapping period is limited by the collision with background gas. The potential depth of a RF trapped ion is much larger than the kinetic energy of the background gas and the ions are trapped in a small area under the irradiation of the probe electromagnetic wave for a long period. RF trapped ions are more advantageous than the optical lattice clocks to observe the narrow linewidth. Operation of an ion trap can be continued for a long period; we can take a long average time for measurement of the transition frequencies of ions.

The transition frequencies of ions are higher than those of neutral atoms with the same electron structure; for example, the $^1S_0$–$^3P_0$ transition frequencies of alkali earth atoms are in the optical region while alkali earth-like ions are in the ultra-violet region. This is because the Coulomb interaction force between the nucleus and electron in ions is stronger than that in neutral atoms. The precision measurement of transition frequency is more advantageous for higher transition frequencies because of the finer scale of time. Therefore, the transition frequencies of ions are in principle more advantageous for precision measurement.

5.4 Precision measurement of hyperfine transition frequencies of alkali-like ions

Atomic clocks were first developed using the transition frequencies in the microwave region, because it is the highest frequency region which could be directly measured using a frequency counter. For the precision measurement of ionic transition frequencies in the microwave region, ions are trapped in a linear trapping apparatus. A traveling microwave propagates in the direction perpendicular to the linear electrode, as shown in figure 5.3. The amplitude of the secular motion perpendicular to the linear electrode is also much less than the wavelength of the microwave at room temperature, therefore, the first order Doppler effect observed as the sideband (section 2.4.2) is negligibly small.

The measurement of the hyperfine transition frequency has been performed with the $^2S_{1/2}$ $(F,m_F) = (0,0) - (1,0)$ transition frequencies of $^{199}Hg^+$, $^{171}Yb^+$, and $^{113}Cd^+$

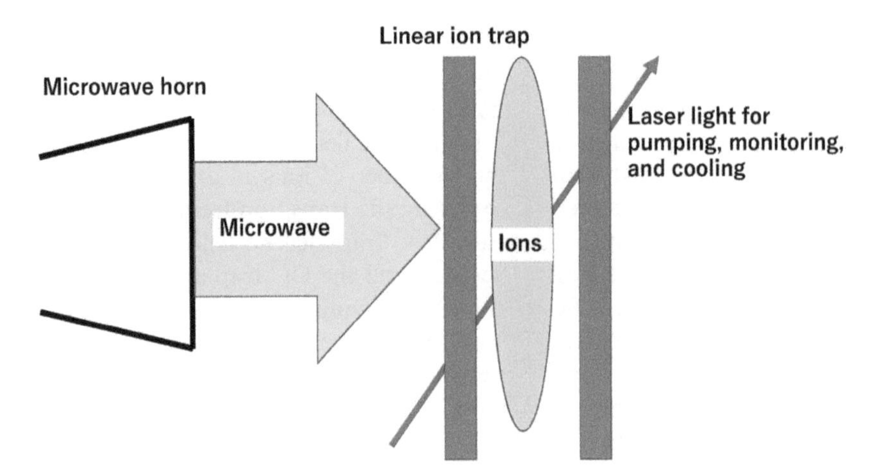

Figure 5.3. Apparatus to observe the microwave transition of ions.

ions, where F is the hyperfine state and m_F is the component of F parallel to the magnetic field [6–11]. This transition is advantageous to measure with a low systematic uncertainty because (1) the ion in the S state is free from the electric quadrupole shift, (2) the Stark shift is small because the energy shifts in both hyperfine states are almost equal, and (3) the Zeeman shift is small because the $m_F = 0$ state is free from the linear Zeeman shift.

The measurement procedure is as follows:

(a) Inducing the $^2S_{1/2}$ $F = 1 \rightarrow {}^2P_{1/2}$ $F = 0$ or 1 transition using a laser light, the ions are pumped to the $^2S_{1/2}$ $F = 0$ state.

(b) Microwave with the frequency of ν_M is irradiated.

(c) Monitor the population in the $^2S_{1/2}$ $F = 1$ state observing the fluorescence inducing the $^2S_{1/2}$ $F = 1 \rightarrow {}^2P_{1/2}$ transition. The fluorescence signal with high intensity is obtained using a laser light, which is resonant with the $^2S_{1/2}$ $F = 1 \rightarrow {}^2P_{1/2}$ $F = 2$ transition because the spontaneous emission transition from the $^2P_{1/2}$ $F = 2$ state is possible only to the $^2S_{1/2}$ $F = 1$ state and the cycle of the laser excitation and spontaneous emission deexcitation repeats.

Laser light has been used for pumping and monitoring [6–11], but $^{202}Hg^+$ lamp has been also used for the measurement of the $^{199}Hg^+$ transition frequency to develop a compact atomic clock (for example, to be used in a satellite) [12]. Performing laser cooling using the $^2S_{1/2}$ $F = 1$ and $0 \rightarrow {}^2P_{1/2}$ $F = 0$ or 1 transition before the measurement procedure, the fractional quadratic Doppler shift is reduced from the order 10^{-13} down to 10^{-17}. For $^{171}Yb^+$ ion, a repump laser resonant to the $^2D_{3/2} \rightarrow {}^2P_{1/2}$ transition is also required. The repump laser is also turned off at procedure (b). When less than 20 ions are laser cooled, ions form a string crystal. Using ions in a string crystal, the fractional Stark shift induced by the trap electric field is below 10^{-17}. To proceed to step (a) after laser cooling, the cooling laser light resonant to the $^2S_{1/2}$ $F = 0 \rightarrow {}^2P_{1/2}$ transition is turned off. Figure 5.4 shows the measurement procedure including laser cooling.

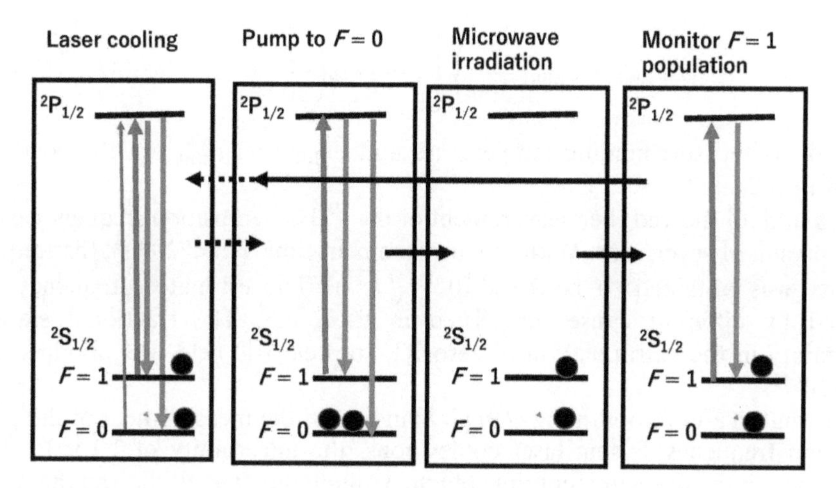

Figure 5.4. The experimental procedure including laser cooling. The red arrows show the optical excitation, and the blue arrows show the spontaneous emission transition.

Table 5.1. The $^2S_{1/2}$ (F,m_F) = (0,0) – (1,0) transition frequencies ν_M and the coefficients of the quadratic Zeeman shift c_{QZ}.

Ion	ν_M (Hz)	c_{QZ} (Hz G^{-2})
$^{199}Hg^+$	40 507 347 996. 841 59 [6]	97.2
$^{171}Yb^+$	12 642 812 118. 4682 [8]	311
$^{113}Cd^+$	15 199 862 856. 633 99 [11]	258

The transition frequencies ν_M and coefficients of the quadratic Zeeman shift c_{QZ} are listed in table 5.1.

The quadratic Zeeman shift $\Delta\nu_{QZ}$ induced by a DC magnetic field is corrected by monitoring the magnetic field also measuring the $\Delta m_F = \pm 1$ transition frequencies $\nu_{M\pm}$ using the approximation

$$\Delta\nu_{QZ} = \frac{(\nu_{M+} - \nu_{M-})^2}{2\nu_M}. \tag{5.4.1}$$

The Zeeman shift is also induced by the AC magnetic field of the stray field (60 Hz) and that induced by the current to give the RF trap electric field, which cannot be eliminated with this method. The uncertainty of the quadratic Zeeman shift is estimated by

$$\delta\left[\Delta\nu_{QZ}\right] = \frac{(\nu_+ - \nu_-)}{\nu_M}\delta[(\nu_+ - \nu_-)]. \tag{5.4.2}$$

There is also a Stark shift with the interaction with blackbody radiation, which is approximately given by a formula,

$$\Delta\nu_{\text{BBR}} = c_{\text{BBR}}\left(\frac{T_s}{300}\right)^4\left[1 + \varepsilon_{\text{BBR}}\left(\frac{T_s}{300}\right)^2\right], \tag{5.4.3}$$

where T_s is the surrounding temperature and c_{BBR} and ε_{BBR} are the coefficients shown in table 5.2.

Berkland [6] showed the measurement of the $^{199}\text{Hg}^+$ transition frequency using a string crystal of seven ions. With the measurement time T_a (<7200 s), the frequency stability was obtained to be $3.3 \times 10^{-13}/\sqrt{T_a(\text{s})}$. The estimated frequency shifts induced by different causes are listed in table 5.3. The fractional statistical uncertainty in the extrapolation of zero RF trap electric field was obtained to be 3.2×10^{-15}.

Warrington and Phoonthong *et al* [7, 8] indicated the measurement of the $^{171}\text{Yb}^+$ transition frequency. Using laser cooled ions, the uncertainty of 3.3×10^{-14} was obtained with the measurement time of 6 h. Mulholland *et al* [9] showed the stability of $3.6 \times 10^{-12}/\sqrt{T_a(\text{s})}$ with the averaging time between 30 s and 1500 s and the prospect to reduce the measurement uncertainty down to below 10^{-14}. The systematic uncertainties caused by several shifts are listed in table 5.4.

Tanaka *et al* [10] reported the measurement of the $^{113}\text{Cd}^+$ transition frequency using ions at room temperature with the measurement uncertainty of 1.4×10^{-10}. Using laser cooled ions, the frequency stability of $4.2 \times 10^{-13}/\sqrt{T_a(\text{s})}$ and the measurement uncertainty of 1.8×10^{-14} was obtained with the measurement time of 5 h [11]. The uncertainty estimated in [11] was mainly dominated by the quadratic Doppler shift, because the kinetic energy was 643 mK.

Table 5.2. The coefficients of the Stark shift induced by blackbody radiation used in equation (5.4.3) [6–11].

Ion	c_{BBR} (Hz)	ε_{BBR}
$^{199}\text{Hg}^+$	-4.01×10^{-6}	
$^{171}\text{Yb}^+$	-1.2×10^{-5}	2×10^{-3}
$^{113}\text{Cd}^+$	-2.76×10^{-6}	

Table 5.3. Estimated frequency shifts in the $^{199}\text{Hg}^+$ $^2\text{S}_{1/2}$ $(F,m_F) = (0,0) \rightarrow (1,0)$ transition frequency [6].

	Fractional shift	Fractional uncertainty
Quadratic Zeeman shift (DC)	2×10^{-14}	1.4×10^{-15}
Quadratic Zeeman shift (60 Hz)	$<2 \times 10^{-20}$	$<2 \times 10^{-20}$
Quadratic Zeeman shift (RF trap current)	5×10^{-15}	3.2×10^{-15}
Blackbody radiation (300 K) Stark shift	$<1.0 \times 10^{-16}$	$<1.0 \times 10^{-16}$
Blackbody radiation (300 K) Zeeman shift	$<1.3 \times 10^{-17}$	$<1.3 \times 10^{-17}$
Stark shift by laser light (194 nm)	$<3 \times 10^{-16}$	$<3 \times 10^{-16}$
Stark shift by trap electric field	$<2 \times 10^{-18}$	$<2 \times 10^{-18}$
Quadratic Doppler shift	$<3 \times 10^{-17}$	$<3 \times 10^{-17}$

Table 5.4. Estimated frequency shifts in the $^{171}Yb^{+}$ $^{2}S_{1/2}$ $(F,m_F) = (0,0) \rightarrow (1,0)$ transition frequency [9].

	Fractional uncertainty
Quadratic Zeeman shift (AC)	$<1.0 \times 10^{-14}$
Blackbody radiation (300 K) Stark shift	$<1.0 \times 10^{-16}$
Stark shift by trap electric field	$<1.0 \times 10^{-15}$
Quadratic Doppler shift	$<1.0 \times 10^{-15}$

Comparing three species of ions, the lowest measurement uncertainty was obtained with the $^{199}Hg^{+}$ transition frequency because the transition frequency is the highest and the ratio of the frequency shifts to the transition frequency is the smallest. Note also that the larger hyperfine splitting makes the coefficient of the quadratic Zeeman shift smaller. However, laser cooling of $^{199}Hg^{+}$ ion is technically complicated because a wavelength of 194 nm is required. Cooling lasers for $^{171}Yb^{+}$ and $^{113}Cd^{+}$ ions are obtained using laser diodes.

Using many ions, we can get an ultra-high frequency stability although the measurement uncertainty below 10^{-13} is difficult to obtain. An atomic clock using 10^{6}–10^{7} $^{199}Hg^{+}$ ions in a linear trap is a candidate as a source of high frequency stability for use in a satellite [12]. The dimensions of 1 liter and 1 kg are attained because optical pumping and transition monitoring (figure 5.4) is performed using a $^{202}Hg^{+}$ lamp. The frequency fluctuation with a measurement time longer than 10^{5} s is below 10^{-15}, which is lower than a hydrogen maser.

The standard of time and frequency is currently determined by the hyperfine transition frequency of neutral Cs atom (9 192 631 770 Hz). Cs atomic clocks with a fountain of cold atoms were constructed in France, Germany, US, UK, Italy, Japan, China, etc. A fractional uncertainty of the order 10^{-16} was obtained by several groups [13]. To our knowledge, a measurement uncertainty below 10^{-15} has never been attained with the microwave transition frequencies of ions. This is mainly because the interest of experimentalists has been shifted to the optical transition frequencies.

5.5 Precision measurement of the optical transition frequencies of atomic ions

The attainable accuracy of atomic clocks based on optical transition is expected to be higher than that for microwave transitions because the time unit of 1 s can be divided into finer scales over five orders. However, optical frequencies cannot be measured directly using a frequency counter. The transition spectrums of atoms or molecules have been observed using laser light since 1970. The frequencies of laser lights were measured using wavelength meters and the fractional uncertainties were of the order of 10^{-7}.

At the beginning of the 21st century, the frequency comb system was developed [14]. A frequency comb is a laser with the simultaneous oscillation of multi-frequency components, which are resonant to the cavity with the length of L_c. The resonant frequencies are given by

$$\nu(N_m) = N_m\nu_{\text{rep}} \quad \nu_{\text{rep}} = \frac{c}{2L_c} \qquad N_m\text{: integer.} \tag{5.5.1}$$

Equation (5.5.1) is in reality not perfectly accurate, because the laser medium has a certain refractive index, and the optical length of the cavity has a slight dependence on the frequency. Changing $L_c \to L_c[1 - \nu_{\text{ceo}}/\nu(N_m)]$, equation (5.5.1) is rewritten as

$$\nu(N_m) = N_m\nu_{\text{rep}} + \nu_{\text{ceo}}. \tag{5.5.2}$$

When the frequency components $\nu(N_m)$ with $N_{\text{min}} \leqslant N_m \leqslant N_{\text{max}}$ oscillate keeping the constant phase relation, the output is observed as a pulse with the repetition rate of $\nu_{\text{rep}}(1/\nu_{\text{rep}}$ is the round trip period of light) and the temporal width of $1/\left[\nu_{\text{rep}}(N_{\text{max}} - N_{\text{min}})\right]$. The frequency components are obtained by measuring ν_{rep} using a frequency counter. The offset frequency ν_{ceo} is obtained from the beat frequency between the second harmonic wave $2 \times \nu(N_m)$ and $\nu(2N_m)$. The frequency of a laser light $\nu_L$is measured using the beat frequency ν_{beat} with one frequency component from the relation of $\nu_L = \nu(N_m) \pm \nu_{\text{beat}}$. Measurement with different repletion rate might be required to distinguish between $\nu(N_m) + \nu_{\text{beat}}$ and $\nu(N_m) - \nu_{\text{beat}}$ or to determine the value of N_m.

Since the development of the frequency comb, the precision measurement of the atomic transition frequencies in the optical region have became one of the hottest subjects for physics researchers. In 1982, Dehmelt gave the first proposal of the precision measurement of transition frequencies of ions in the optical region [15]. The measurement should be performed with a E1 forbidden transition to a metastable state, so that the natural linewidth is narrower than 10 Hz. E1 allowed transitions are not used as a standard frequency because the natural spectrum linewidth is broadened to several MHz. Note also that the spectrum of the E1 allowed transition is distorted because there is a cooling effect with the frequency lower than the transition frequency, while there is a heating effect at higher frequency (see figure 2.7). A single ion trapped at the center of a three-dimensional RF trap (electric field is zero) is used for the measurement of the transition frequency. The secular motion amplitude is reduced by laser cooling to one order smaller than the wavelength of the probe laser light (secular motion energy <1 mK), so that (1) the sideband transition rates (section 2.4.2) are negligibly small in comparison with the carrier transition rate, (2) the Stark shift induced by the trap electric field is reduced to below 10^{-18}, and (3) the quadratic Doppler shift is reduced to below 10^{-17}.

Except for the transition between states without electric quadrupole moment, the elimination of the electric quadrupole shift is required as shown in section 5.5.1.

The measurement procedure starts with the laser cooling, with which fluorescence is observed. After cooling, the cooling laser is turned off and the probe laser is switched on. A mechanical shutter and acoustic optical modulator are often simultaneously used to block the cooling laser perfectly, because a significant Stark shift is induced when the blocking is not perfect. To monitor the transitions induced by the probe beam, the cooling lasers are turned on again (figure 5.5). When the transition to a metastable state (clock transition) is induced by the probe beam, fluorescence is also not observed when the cooling laser light is irradiated. Without

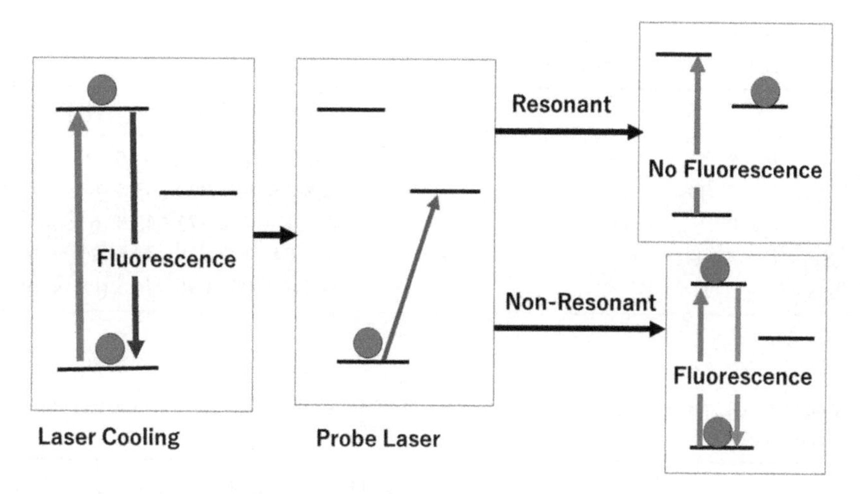

Figure 5.5. Processes in the transition of a single ion. Laser cooling is performed using a E1 allowed transition. Probe laser is irradiated after the cooling laser is turned off. After a while, the cooling laser is irradiated again. If the probe laser induced the transition, no fluorescence is observed. Without the transition induced by the probe laser, fluorescence is observed. Reproduced from [1].

the clock transition, fluorescence is observed when the cooling laser is irradiated. This measurement procedure is applicable for all ions for which laser cooling and fluorescence observations are both possible.

Less than ten ions in a linear trap (two-dimensional RF trap and one-dimensional DC trap) form a string crystal with ultra-low kinetic energy. The electric field is negligibly small for ions in a string crystal. Therefore, measurement with multiple ions is possible when a linear trap is used. A linear trap is useful to measure the transition frequencies of ions which cannot be laser cooled, because they can be cooled by the Coulomb interaction with co-trapped laser cooled ions (called sympathetic cooling). It is also possible to monitor the state of ions without direct observation of the fluorescence from a cycle transition because the information of the state of an ion can be transformed to that of co-trapped ion after making an entangled state (section 3.3). By monitoring the state of a co-trapped ion, the state of the target ion is determined.

The electric quadrupole shift is significant for the measurement using linear trap, because of the large DC electric field gradient. Therefore, the precision measurement using a linear trap has been performed mainly for the transition frequencies between states where the electric quadrupole moment is zero.

5.5.1 Measurement with alkali-like ions

Experimental studies of alkali-like ions have been performed by different research groups. The $^2S_{1/2}$–$^2D_{5/2}$, $^2D_{3/2}$ transition frequencies are good candidates for precision measurement, because the natural spectrum linewidth is narrow (order of 1 Hz). The transition frequencies recommended at Consultative Committee for Time and frequency (CCTF) 2017 are listed in table 5.5 [16].

Table 5.5. Values of transition frequencies of alkali-like ions recommended at CCTF2017. The recommended fractional uncertainties are different from those indicated in the text [16].

	Transition frequencies (Hz)
$^{199}Hg^+$ $^2S_{1/2}$ $F = 0$–$^2D_{5/2}$ $F = 2$	1 064 721 609 899 145.30 (1.9×10^{-15})
$^{171}Yb^+$ $^2S_{1/2}$ $F = 0$–$^2D_{/3/2}$ $F = 2$	688 358 979 309 308.3 (6×10^{-16})
$^{171}Yb^+$ $^2S_{1/2}$ $F = 0$–$^2F_{7/2}$ $F = 3$	642 121 496 772 645.0 (6×10^{-16})
$^{40}Ca^+$ $^2S_{1/2}$–$^2D_{5/2}$	411 042 129 776 399.8 (2.4×10^{-15})
$^{88}Sr^+$ $^2S_{1/2}$–$^2D_{5/2}$	444 779 044 095 486.5 (1.5×10^{-15})

Table 5.6. DC polarizability α_p and the coefficient of the Stark shift induced by the blackbody radiation c_{BBR}.

	α_p (mHz V^{-2} cm^{-2})	c_{BBR} (Hz)
$^{199}Hg^+$ $^2S_{1/2}$ $F = 0$–$^2D_{5/2}$ $F = 2$	2.28	−0.072
$^{171}Yb^+$ $^2S_{1/2}$ $F = 0$–$^2D_{/3/2}$ $F = 2$	11.2	−0.36
$^{171}Yb^+$ $^2S_{1/2}$ $F = 0$–$^2F_{7/2}$ $F = 3$	1.4	−0.045
$^{40}Ca^+$ $^2S_{1/2}$–$^2D_{5/2}$	−10.7	0.4
$^{88}Sr^+$ $^2S_{1/2}$–$^2D_{5/2}$	−7.3	0.3

The Stark shift by the DC electric field E_{DC} is given by

$$\Delta\nu_{\mathrm{S-DC}} = -\frac{\alpha_p}{2} E_{\mathrm{DC}}^2 \tag{5.5.3}$$

where α_p is the polarizability, listed in table 5.6. The Stark shift induced by the RF trap electric field is estimated using equation (5.5.3), because the frequency of the RF trap electric field is much lower than the transition frequencies between electric dipole coupled states (appendix B). The Stark shift induced by the blackbody radiation with the surrounding temperature T_s (K) is given by

$$\Delta\nu_{\mathrm{BBR}} = c_{\mathrm{BBR}} \left(\frac{T_s(\mathrm{K})}{300} \right)^4 \tag{5.5.4}$$

where c_{BBR} is listed in table 5.6.

Even isotopes of alkali-like ions have no nuclear spin, therefore, the energy structure is simple and laser cooling is easy. On the other hand, m_F ($= m_J$) are half integers and the $m_F = 0$ state (free from the linear Zeeman shift) does not exist. The linear Zeeman energy shift at each state is proportional to m_F and there is a significant linear Zeeman shift at each $m_F = m_F' \to m_F''$ transition frequencies. For example, the linear Zeeman coefficient of the $^2S_{1/2}$ $m_F = \pm 1/2 \to$ $^2D_{5/2}$ $m_F = \pm 5/2$ is ± 2.8 MHz G^{-1}. It was thought up to 2004 that the odd isotopes were better candidates for precision measurement, because F is an integer, and the $m_F = 0$ state (free from linear Zeeman shift) exists [17]. Table 5.7 lists the coefficients of the quadratic Zeeman shifts in several transition frequencies with $m_F = 0 \to 0$.

Table 5.7. The coefficient of the quadratic Zeeman shift in the $m_F = 0 \rightarrow 0$ transitions.

Transitions	Coefficient of quadratic Zeeman shift (Hz G^{-2})
$^{199}Hg^+$ $^2S_{1/2}$ $F = 0$–$^2D_{5/2}$ $F = 2$	−189
$^{171}Yb^+$ $^2S_{1/2}$ $F = 0$–$^2D_{3/2}$ $F = 2$	520
$^{171}Yb^+$ $^2S_{1/2}$ $F = 0$–$^2F_{7/2}$ $F = 3$	−20.2
$^{43}Ca^+$ $^2S_{1/2}$ $F = 4$–$^2D_{5/2}$ $F = 6$	−89 900
$^{87}Sr^+$ $^2S_{1/2}$ $F = 5$–$^2D_{5/2}$ $F = 7$	64 000

For the $^{199}Hg^+$ and $^{171}Yb^+$ transition frequencies, the fractional quadratic Zeeman shifts below 10^{-16} with the magnetic field of 10 mG. For the $^{43}Ca^+$ and $^{87}Sr^+$ transition frequencies, the magnetic field is required to be below 2 mG to reduce the fractional quadratic Zeeman shift to below 10^{-15}. There is also technical merit to measuring the $^{199}Hg^+$ and $^{171}Yb^+$ transition frequencies, that the localization to the $m_F = 0$ state is attained just by pumping to the $F = 0$ state.

The measurement of the $^{199}Hg^+$ transition frequency was performed first [18]. The electric quadrupole moment exists in the D state, and the electric quadrupole shift is one component to give the limit of the measurement uncertainty. The electric quadrupole shift $\Delta\nu_{eQ}$ is proportional to $[3\cos(\beta_2)^2 - 1][3m_F^2 - F(F+1)]$, where β_e is the angle between the directions of the magnetic field and the electric field gradient. With this experiment, the electric quadrupole shift was eliminated by averaging the results applying the magnetic field in three orthogonal directions. It is eliminated also by averaging the transition frequencies with $-F \leqslant m_F \leqslant F$. The experiment with $^{199}Hg^+$ ion is required to be performed with a cryogenic chamber using liquid He to suppress the Hg vapor pressure, therefore the Stark shift induced by the blackbody radiation is below 10^{-20}. The frequency stability of 7×10^{-15} was measured with the averaging time of 1 s [18]. The measurement uncertainty of 1.9×10^{-17} was estimated from the uncertainties of shifts listed in table 5.8 [19].

The measurement of the $^2S_{1/2}$ $F = 0 \rightarrow$ $^2D_{3/2}$ $F = 2$ transition frequency was started by a group in Germany, while the measurement of the $^2S_{1/2}$ $F = 0 \rightarrow$ $^2F_{7/2}$ $F = 3$ transition frequencies was started by a group in UK [20, 21]. The S–F transition frequency seems to be more advantageous than the S–D transition frequency for precision measurement because of narrower spectrum linewidth and small Stark (trap electric field, blackbody radiation), Zeeman, and electric quadrupole shifts (see table 5.9). However, high intensity of the probe laser is required to induce the S–F transition. A Stark shift of 100 Hz is induced by the probe laser. This problem was overcome by the development of a method called the hyper-Ramsey spectroscopy, with which the Stark shift induced by the probe laser is suppressed [22]. Both transition frequencies are now measured by both groups. The measurement uncertainty of the S–F transition frequency is estimated to be 3×10^{-18}, while it is 1×10^{-16} for the S–D transition frequency [23]. The estimation of Stark, Zeeman, and electric quadrupole shifts in the S–D transition frequency is now used to monitor the uncertainties of electric field, magnetic field, and electric field gradient, which are used to estimate the uncertainty of shifts in the S–F transition frequency.

Table 5.8. Uncertainties of shifts in the $^{199}Hg^{+}$ $^{2}S_{1/2}$ $F = 0 \rightarrow$ $^{2}D_{5/2}$ $F = 2$ transition frequency [19].

Cause of the frequency shift	Fractional uncertainty
Quadratic Doppler shift(micromotion)	4×10^{-18}
Quadratic Doppler shift (secular motion)	3×10^{-18}
Quadratic Zeeman shift (DC)	5×10^{-18}
Quadratic Zeeman shift (AC)	10×10^{-18}
Electric quadrupole shift	10×10^{-18}

Table 5.9. Uncertainties of shifts in the $^{171}Yb^{+}$ $^{2}S_{1/2}$ $F = 0 \rightarrow$ $^{2}D_{3/2}$ $F = 2$ (S–D) [20] and the $^{2}S_{1/2}$ $F = 0 \rightarrow$ $^{2}D_{7/2}$ $F = 3$ (S–F) [21] transition frequencies.

Cause of the frequency shift	Fractional uncertainty (S–D)	Fractional uncertainty (S–F)
Quadratic Doppler shift	3×10^{-18}	2.1×10^{-18}
Blackbody radiation shift	102×10^{-18}	1.8×10^{-18}
Probe laser related shift		1.1×10^{-18}
Quadratic Zeeman shift	7×10^{-18}	0.6×10^{-18}
Stark shift by trap electric field	7×10^{-18}	0.5×10^{-18}
Electric quadrupole shift	14×10^{-18}	0.5×10^{-18}

The measurement of the $^{2}S_{1/2} \rightarrow$ $^{2}D_{5/2}$ transition frequencies of $^{88}Sr^{+}$ and $^{40}Ca^{+}$ ions have been performed since 2004, eliminating the linear Zeeman shift by averaging the $m_J = m_J' \rightarrow m_J''$"and $m_J = -m_J' \rightarrow -m_J''$"(for even isotopes of alkali-like ions, $m_F = m_J$ because the nuclear spin is zero) [24–27]. By averaging the $m_J = \pm 1/2 \rightarrow \pm 1/2, \pm 1/2 \rightarrow \pm 3/2$, and $\pm 1/2 \rightarrow \pm 5/2$ transition frequencies, not only the linear Zeeman shift but also the electric quadrupole shift is eliminated. The AC magnetic field (induced by the commercial electric current) makes the spectrum linewidth broadened by the modulated linear Zeeman shift. Suppressing the AC magnetic field using a magnetic shield, the spectrum linewidth was reduced. The magnetic shield suppressed the DC magnetic field also to below 10 mG and the quadratic Zeeman shift (coupling between the $^{2}D_{5/2}$ and $^{2}D_{3/2}$ states) was reduced to below 0.1 mHz. Dube *et al* [25] showed the frequency stability of 3×10^{-15} with the average time of 1 s and the systematic fractional uncertainty of 1.5×10^{-17} with the $^{88}Sr^{+}$ transition frequency. Huang *et al* [27] indicate the comparison between two $^{40}Ca^{+}$ clocks with the fractional difference of 3.2×10^{-17} with the uncertainty of 5.5×10^{-17}. The fractional stability was 7×10^{-17} with the average time of 20 000 s. The use of $^{88}Sr^{+}$ and $^{40}Ca^{+}$ transition frequencies has a technical advantage, because of the negative polarizability (positive DC Stark shift). When the ion is trapped at the position shifted from the center (RF trap electric field is zero), there is non-zero trap electric field of E_t. There is a micromotion with the amplitude of $eE_t/m\Omega^2$ and frequency of $\Omega/2\pi$ (see equation (1.5.2)). The Stark shift $\Delta\nu_{S-DC} = -\alpha_p E_t^2/2(>0)$ and the quadratic Doppler shift $-(eE_t/m\Omega c)^2\nu_c/2(<0)$ cancel each other with a magic RF trap frequency [28]

$$\frac{\Omega_{mg}}{2\pi} = \frac{e}{2mc}\sqrt{-\frac{\nu_c}{\alpha_p}}, \tag{5.5.5}$$

which was obtained to be 14.4 MHz and 24.9 MHz for the $^{88}Sr^+$ and $^{40}Ca^+$ transition frequencies, respectively. Trapping with the magic RF trap frequency, measurement of transition frequency can be performed using multiple ions. There is also a proposal to obtain a high frequency stability using $^{40}Ca^+$ ions in a circle crystal in a multipole trap [29].

5.5.2 Measurement with alkali earth-like ions

The $^1S_0 \rightarrow {}^3P_0$ transition frequencies of alkali earth-like ions are advantageous for precision measurement because they are free from the electric quadrupole shift (because of $J = 0$ at both states). The $J = 0 \rightarrow 0$ transition is in principle forbidden, as shown in section 2.1. However, there is a slight mixture between the $^3P_0$ and $^3P_1$ states when there is non-zero nuclear spin (a magnetic field is applied from the nucleus) and the transition is possible (see section 2.1). The transition frequencies of $^{119}In^+$ and $^{27}Al^+$ ions recommended at CCFH2017 are listed in table 5.10 [16]. Because of this state mixture, there is a linear Zeeman shift with the coefficients of several kHz G^{-1}, which is three orders smaller than that for the even isotope if alkali-like ions. Averaging the $m_F = \pm m_F' \rightarrow \pm m_F''$ transition frequencies, the linear Zeeman shift is eliminated. The magnetic field is estimated from the difference of both transition frequencies, which is used to eliminate the quadratic Zeeman shift induced by the interaction between $^3P_0$ and $^3P_1$ states, whose coefficients are listed in table 5.10.

The Stark shift in this transition frequency is smaller than that for alkali-like ions because the energy gaps between different states in alkali earth-like ions are much larger than those for alkali-like ions and the interaction with other states are much weaker. Values of polarizabilities and the blackbody radiation shift with a temperature of 300 K are listed in table 5.10.

However, there are technical difficulties in treating alkali earth-like ions. In principle, laser cooling and fluorescence detection are possible using the $^1S_0 \rightarrow {}^1P_1$ transition. However, laser light with a very short wavelength (167 nm for $^{27}Al^+$, 159 nm for $^{119}In^+$) is required, which is not realistic.

Table 5.10. $^1S_0 \rightarrow {}^3P_0$ transition frequencies of $^{119}In^+$ and $^{27}Al^+$ ions recommended at CCTF2017 [16]. The coefficient of the quadratic Zeeman shift, polarizability for DC electric field, and the Stark shift induced by the blackbody radiation with the temperature of 300 K are also listed.

	$^{119}In^+$	$^{27}Al^+$
Transition frequency (Hz)	1 267 402 452 899 920	1 121 015 393 207 857.3
Quadratic Zeeman shift (Hz/G^2)	−0.041	−0.719 88
Polarizability (mHz V^{-2} cm^{-2})	0.99	0.26
Blackbody radiation 300 K (Hz)	−0.034	−0.009

The measurement of the $^{119}In^+$ transition frequency was first performed using the $^1S_0 \rightarrow {}^3P_1$ transition for laser cooling and fluorescence detection [30]. The rate of the spontaneous emission transition is $2\pi \times 360$ kHz, with which laser cooling is possible (although the cooling force is weaker than using the S–P transition). There is also the merit that the Doppler cooling limit is lower than laser cooling of alkali-like ions using S–P transition. The fluorescence detection rate cannot be high, but a high S/N ratio can be obtained suppressing the detection noise from the blackbody radiation using a photodetector which is sensitive only for photons in the ultra-violet (UV) region.

The $^1S_0 \rightarrow {}^3P_0$ transition frequency can be measured using a linear trap, because it is free from the electric quadrupole shift. The group in National Institute of Information and Communications Technology (NICT, Japan) indicates the measurement using $^{119}In^+$ ion in a linear trap [31]. $^{40}Ca^+$ ion is co-trapped, and laser cooled, then also the $^{119}In^+$ ion is sympathetically cooled. With this method, the measurement uncertainty of 7.7×10^{-16} in comparison with the $^1S_0 \rightarrow {}^3P_0$ transition frequency of ^{87}Sr atoms in an optical lattice. The measurement uncertainty was dominated by the quadratic Doppler shift and the statistical uncertainty. The statistical measurement uncertainty can be reduced using many $^{119}In^+$ ions. The group in Physikalisches Technische Bundesanstalt (PTB, Germany) compared the systematic uncertainties when $^{119}In^+$ ions were sympathetically cooled with Yb^+ ions and when they were cooled by direct laser cooling using the $^1S_0 \rightarrow {}^3P_1$ transition, as shown in table 5.11 [32]. From this result, it seems preferable to use the direct laser cooling, because of the lower Doppler cooling limit.

It is difficult to use the same experiment to measure the $^{27}Al^+$ transition frequency, because the $^1S_0 \rightarrow {}^3P_1$ transition rate is lower than that for $^{119}In^+$ ion. The measurement of the $^{27}Al^+$ transition frequency has been performed by sympathetic cooling with co-trapped alkali-like ions ($^9Be^+$, $^{25}Mg^+$, $^{40}Ca^+$) and the state detection by the quantum logic method (make an entangled state with the co-trapped ion and monitor the state of co-trapped ion, as shown in section 3.3) [19, 33–35]. The fractional measurement uncertainty of 2.3×10^{-17} was attained in 2007, which was dominated by the quadratic Doppler shift [19]. The kinetic energy was reduced by sideband Raman cooling (section 2.4) for all motion modes [34] and the

Table 5.11. The fractional systematic measurement uncertainty of the $^{119}In^+$ $^1S_0 \rightarrow {}^3P_0$ transition frequency when $^{119}In^+$ ions were sympathetically cooled with Yb^+ ions and when they were cooled by direct laser cooling using the $^1S_0 \rightarrow {}^3P_1$ transition [32].

	Sympathetic cooling	Direct laser cooling
Quadratic Doppler shift (thermal)	4×10^{-19}	0.4×10^{-19}
Quadratic Doppler shift (micromotion)	0.6×10^{-19}	0.8×10^{-19}
Stark shift (trap electric field)	0.1×10^{-19}	0.1×10^{-19}
Electric quadrupole shift	$<10^{-21}$	$<10^{-21}$
Blackbody radiation	0.54×10^{-19}	0.15×10^{-19}
Total	4.1×10^{-19}	0.9×10^{-19}

Table 5.12. The fractional systematic measurement uncertainty of the $^{27}Al^{+}$ $^1S_0 \rightarrow {}^3P_0$ transition frequency [34].

	Fractional uncertainty
Quadratic Doppler shift (micromotion)	5.9×10^{-19}
Quadratic Doppler shift (secular motion)	2.9×10^{-19}
Quadratic Zeeman shift	3.7×10^{-19}
Stark shift (Blackbody radiation)	4.2×10^{-19}
Stark shift (probe laser)	2.0×10^{-19}

measurement uncertainty of 9.4×10^{-19} and the stability of $1.2 \times 10^{-15}/\sqrt{T_a(s)}$ were attained [35]. Table 5.12 lists the fractional uncertainties of frequency shifts induced by several causes.

The energy structure of $^{171}Yb^{2+}$ ion is alkali earth-like [36]. The clock transition is 1.35×10^{15} Hz (wavelength of 220 nm). The cooling should be performed by sympathetic cooling with laser cooled Yb^{+} ions. The $^1S_0 \rightarrow {}^3P_1$ transition is induced using a laser with the wavelength of 252 nm and the spontaneous emission rate is s $2\pi \times 690$ kHz. Therefore, we can observe the transition by direct observation of fluorescence from the $^1S_0 \leftrightarrow {}^3P_1$ cycle transition of $^{171}Yb^{2+}$ ion. The measurement uncertainty of 10^{-18} also seems to be attainable with this transition.

5.5.3 Measurement with highly charged ions

The transition frequencies of highly charged ions X^{q+} are also good candidates for precision measurement. With highly charged ions, the interaction between nucleus and electron is much stronger than that for atoms with the same number of electrons. Therefore, the transition frequencies of highly charged ions are much higher than the same transition frequencies of neutral atoms or single charged ions with the same electron number; for example, the hyperfine transition frequencies are in the optical region with highly charged ions. The quadratic Stark and Zeeman shifts in the transition frequency are much less than for neutral atoms because energy gaps between the different electron states are much larger than those for neutral atoms.

There is a technical difficulty in controlling highly charged ions because the initial kinetic energy is of the order 1 MK. Laser cooling of a highly charged ion is not realistic because the electronic transition cannot be induced by laser light operating in the optical region. The cooling should be performed by sympathetic cooling with a laser cooled single charged ion. For co-trapping with a uniform frequency and amplitude of trap electric field, the mass of the co-trapped ion should be close to m/q. On the other hand, the sympathetic cooling effect is small because of the large mass ratio with the co-trapped ion. Therefore, sympathetic cooling should be performed with many co-trapped ions. Micke *et al* [37] indicated that the $^{40}Ar^{13+}$ E1 $^2P_{1/2} \rightarrow {}^2P_{3/2}$ (E1 forbidden) transition frequency (441 nm) was measured with the uncertainty of 10^{-15}. In this experiment, $^{40}Ar^{13+}$ ion was cooled by interactions with

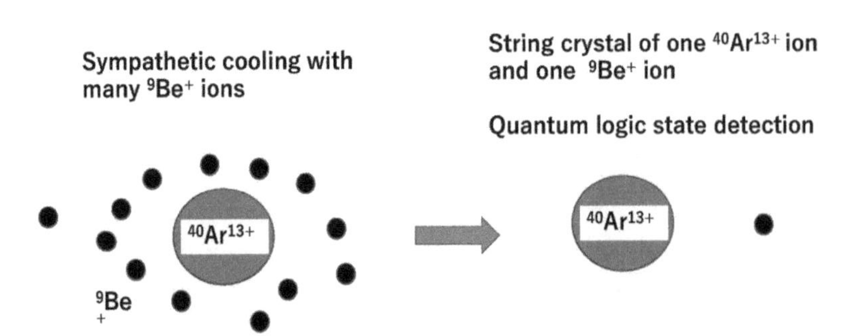

Figure 5.6. Sympathetic cooling of $^{40}Ar^{13+}$ ion with many $^{9}Be^{+}$ ions, and quantum logical state detection with a string crystal of one $^{40}Ar^{13+}$ ion and one $^{9}Be^{+}$ ion.

many Be^{+} ions which were laser cooled (sympathetic cooling) and made a string crystal of one $^{40}Ar^{13+}$ ion and one $^{9}Be^{+}$ (the other $^{9}Be^{+}$ ions are removed) in a linear trap as shown in figure 5.6. After the forbidden optical transition of 441 nm was induced, the state detection was performed with the quantum logical method, making an entangled state with the co-trapped ion (section 3.3).

The experimental treatment of highly charged ions is much more complicated than single charged ions; however, their transition frequencies are good candidates for precision measurement in the future. As the Coulomb interactions between electron and nucleus in highly charged ions are much stronger than those in neutral atoms and single charged ions, the relativistic effect with the electron motion is significant. Therefore, measurement of transition frequencies can contribute significantly to the development of new physics. For example, the transition frequencies of highly charged ions are sensitive to the variation in fine structure constant (see section 5.8) [38].

5.6 Measurement of transition frequencies of molecular ions

The measurement of molecular transitions are also hot topics to developing new physics. However, the molecular energy structure is much more complicated than atomic transition frequencies with vibrational rotational states, as shown in section 2.1. There is also hyperfine structure when nuclear spins of composing atoms are not zero. With the thermal distribution at room temperature, molecules are distributed to several hundred different states. The laser induced transition is caused only by a small fraction of the molecules in the targeting state and the S/N ratio of the molecular spectrum is much smaller than that of the atomic spectrum. Spontaneous emission transition from one state is possible to several vibrational rotational states, therefore, it is difficult to observe the fluorescence from a cycle of laser induced excitation + spontaneous emission deexcitation, which makes it difficult to monitor the state by observing the fluorescence. Laser cooling is also difficult for the same reason.

It would appear to be possible to overcome these difficulties with diatomic molecular ions, whose energy structure is much simpler than multiatomic molecules. It is preferable that it includes an atom with zero nuclear spin to avoid a complicated hyperfine structure. Optical pumping to the vibrational rotational ground state can be performed using a laser with a wide frequency range (for example, optical frequency comb) which induces the excitation to the electric excited state from all vibrational rotational states except for from the vibrational rotational ground state [39]. The kinetic energy of molecular ions in a linear trap is reduced by sympathetic cooling with co-trapped and laser cooled atomic ions. The state detection of a single molecular ion is possible with the quantum logic method, making an entangled state with the co-trapped atomic ion as shown in section 3.3. For previous experiments using many molecular ions, state detection has been performed by state selective photodissociation or the second ionization (not applicable for measurement with the uncertainty below 10^{-15} because the string crystal is not formed with many ions).

5.6.1 Experimental results of the molecular ion spectrum

The transition of the $^{40}CaH^+$ molecular ion from the electronic ground state $X^1\Sigma$ $n_v = 0 J = J''$ to the first electronic excited state $A^1\Sigma$ $n_v = 0\text{–}3 J = J'$ was observed using sympathetic cooling with co-trapped $^{40}Ca^+$ ion and the detection of ions in the electronic excited state by selective photodissociation [40]. Here, n_v is the vibrational quantum number and J is the rotational quantum number ($J = N$ in the $^1\Sigma$ state, where N is the general expression of rotational state). The energy state in both electric energy states is approximately given by (here the unharmonic vibrational term is also considered)

$$\frac{E_{st}}{h} = T_e + \nu_{\text{vib}}\left(n_v + \frac{1}{2}\right) - x\nu_{\text{vib}}\left(n_v + \frac{1}{2}\right)^2 + B_{n_v}J(J + 1). \quad (5.5.6)$$

In the $X^1\Sigma$ state, $T_e = 0$. For the $A^1\Sigma$ state, the following values were experimentally obtained

$T_e = 72\ 660\pm150$ GHz $\nu_{\text{vib}} = 24372\pm180$ GHz $x\nu_{\text{vib}} = 503\pm42$ GHz

$B_0 = 95.3\pm2.4$ GHz $B_1 = 85.7\pm0.90$ GHz $B_2 = 88.1\pm1.5$ GHz $B_3 = 85.4\pm0.90$ GHz.

The vibrational rotational transition frequencies of HD^+ molecular ion were measured as shown in table 5.13 [41–44]. The electronic ground state of HD^+ molecular ion is $X^2\Sigma$ state (electron spin 1/2, electron orbital angular momentum 0).

Table 5.13. Measurements of the HD^+ vibrational rotational transitions frequencies [41–44]. Here, n_v and N are quantum numbers of vibrational and rotational states, respectively.

(n_v, N)	Transition frequency (kHz)
(0,0) → (0,1)	1 314 935 828.0±0.4
(0,0) → (1,1)	58 605 052 000±64
(0,2) → (4,3)	214 978 560 600±500
(0,2) → (8,3)	383 407 177 380±410

The precision measurement of HD^+ transition frequencies is useful to determine fundamental constants comparing the measurements of the transition frequencies with rigorous calculations using proper constants,

For example. the proton-to-electron mass ratio and D nucleus-to-electron mass ratio were determined to be 1836.1512±0.0025 and 3670.4885±0.0101, respectively. There is a technical difficulty observing the HD^+ transitions because of the complicated hyperfine structures (electron spin 1/2, H nuclear spin 1/2 and D nuclear spin 1).

Germann *et al* [45] indicated the observation of the ${}^{14}N_2{}^+ X^2\Sigma(n_v, N) = (0,0) \rightarrow (1,2)$ transition. The ${}^{14}N_2{}^+$ molecular ion was prepared in a selected state by state selective photoionization of ${}^{14}N_2$ molecules shown in figure 1.13. The state detection was performed using the ${}^{14}N_2{}^+(n_v \geqslant 1) + Ar \rightarrow {}^{14}N_2 + Ar^+$ reaction. The ${}^{14}N$ nuclear spin is 1 and the total nuclear spin I can be 0, 1, and 2. From the symmetry of the homonuclear diatomic molecular ion, I can be 0 or 2 with even rotational state and $I = 1$ for the odd rotational states. The electron spin is 1/2 and the fine and hyperfine structure in $N = 0$ and 2 states are given by

$$\begin{aligned}
&I = 0\\
&N = 0, \;\; J = 1/2\\
&N = 2, \;\; J = 5/2, \;\; 3/2\\
&I = 2\\
&N = 0, \;\; J = 1/2, F = 5/2, \;\; F = 3/2\\
&N = 2, \;\; J = 5/2, \;\; F = 9/2, \;\; 7/2, \;\; 5/2, \;\; 3/2, \;\; 1/2\;1/2\\
&\qquad\quad\;\; J = 3/2, \;\; F = 7/2, \;\; 5/2, \;\; 3/2, \;\; 1/2\;1/2
\end{aligned}$$

In [45], transitions satisfying $\Delta N = \Delta J = \Delta F = 2$ were observed, because the transition rates are relatively high. The measured transition frequencies were transition (i) $I = 0\ (n_v, N, J) = (0,0, 1/2) \rightarrow (1,2, 5/2)$ 65 539.831±0.012 GHz

transition (ii) $I = 2(n_v, N, J, F) = (0,0, 1/2,5/2) \rightarrow (1,2, 5/2,9/2)$
65 539.815±0.012 GHz

transition (iii) $I = 2\ (n_v, N, J, F) = (0,0, 1/2,3/2) \rightarrow (1,2, 5/2,7/2)$
65 540.039± 0.012 GHz.

In [45], transitions (i) and (ii) overlap. Using a probe laser with a narrower linewidth, both transitions are expected to be resolved, because the natural linewidth (given by spontaneous emission rate) was estimated to be 11 nHz and the spectrum linewidth is given by the linewidth of the probe laser.

The state detection of the molecular ion succeeded using a string crystal of one ${}^{40}CaH^+$ molecular ion and one ${}^{40}Ca^+$ ion [46]. The $\Delta J = 2$ rotational transition ν_R in the vibrational ground state was observed using a frequency comb (see section 5.5). When the repetition frequency of the frequency comb is an integer division of the transition frequency $\left(\nu_{\text{rep}} = \nu_R / n_R\right)$, a Raman transition is induced by the frequency components $\nu(N_m)$ and $\nu(N_m + n_R)$. There are many pairs of frequency component to induce the Raman transition in a frequency comb, therefore, high transition efficiency can be obtained. The rotational constant of $^{40}CaH^+$ molecular ion in the $X^1\Sigma n_v = 0$ state was obtained to be B_0 = 142 501 777.9±1.7 kHz. The rotational state was detected by the quantum logical method.

5.6.2 Prospect for the precision measurements of transition frequencies of molecular ions

The measurement uncertainties of transition frequencies of molecular ions shown in section 5.6.1 are of the order of 1 ppb. To our knowledge, there has been no report about the measurement of molecular transition frequencies with uncertainties below 10^{-15}.

Here we discuss the attainable systematic uncertainty of molecular transition frequencies. The quadratic Doppler shift has been suppressed by sympathetic cooling with co-trapped laser cooled atomic ions. The gravitational red shift is a common effect for all transition frequencies, which depend only on the gravitational potential of the measuring position. Therefore, the attainable accuracy should be discussed with the Stark, electric quadrupole, and Zeeman shifts. Generally the molecular energy structure is complicated, and there are interactions between different quantum states. Therefore, the estimation of Stark, electric quadrupole, and Zeeman shifts is more difficult than the atomic transition frequencies.

However, note that the transition frequencies are not shifted by the Stark, electric quadrupole, and Zeeman effects when the energy shifts in the upper and lower states are equal. Comparing the vibrational rotational states in the electronic ground state, these shifts depend significantly on the angular momentum quantum numbers (electron spin, molecular rotation, etc), which gives the directional distribution of the wavefunction (molecular figure). For example, the wavefunction of the $N = 0$ state (N: rotational quantum number) is spherical symmetric with any vibrational state. These shifts have very small dependence on the vibrational state, because the change of the vibrational state gives only a slight change of the bond length (a few % with the change of one vibrational state). Therefore, cancellation of the Stark, electric quadrupole, and Zeeman energy shifts at upper and lower states are significant for the vibrational transitions without the change of any angular momentum quantum numbers (pure vibrational transition) (figure 5.7).

The $N = 0 \rightarrow 0$ $\Delta J = \Delta F = \Delta m_F = 0$ vibrational transitions in the $^1\Sigma$ and $^2\Sigma$ states seem to be particularly advantageous for precision measurement because it is free from the electric quadrupole shift because the electric quadrupole moment is zero in the $N = 0$ state. The linear Zeeman shift is also particularly small for the $N = 0 \rightarrow 0$ transition frequency. The Zeeman shift is induced by the electron spin, nuclear spin, and molecular rotation. The linear Zeeman coefficient is dominated by the term given by the electron spin, but this value is the same for any vibrational state and the Zeeman shifts at vibrational ground and excited states cancel each other. The linear Zeeman shift induced by the nuclear spin has a dependence on the vibrational state with the order of 0.1 ppm, which is cause by the change of the magnetic shielding effect by the electron cloud surrounding the nucleus. The linear

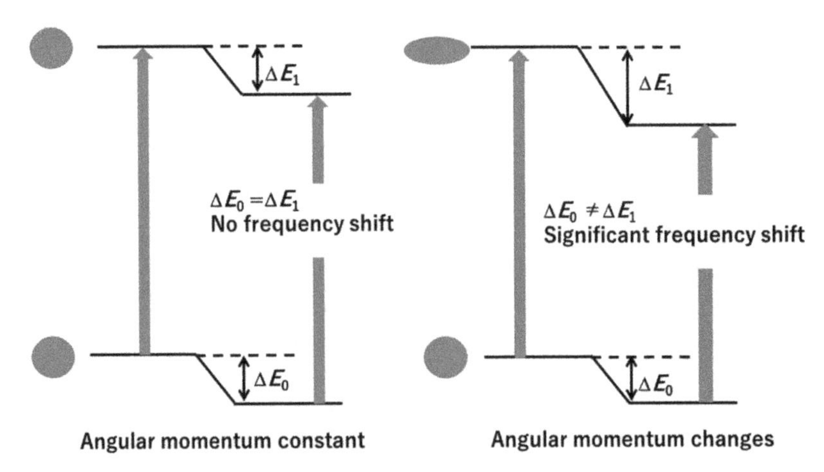

Figure 5.7. Frequency shifts induced by an electric field, electric field gradient or a magnetic field for the upper and lower states in molecular transitions. With the angular momentum quantum numbers remaining constant, the two frequency shifts cancel each other out. With the change of the angular momentum quantum numbers, there is significant shift in the transition frequency. Reproduced from [1].

Zeeman shift induced by the molecular rotation changes by a few % with the change of the vibrational state. The linear Zeeman coefficients in the $N = 0 \rightarrow 0$ pure vibrational transition frequencies are expected to be below 1 mHz/G, while it is order of 1 Hz/G for the $N = 1 \rightarrow 1$ pure vibrational transition frequencies.

5.6.3 The vibrational transition frequencies of diatomic polar molecular ions

First, we estimate the attainable measurement uncertainty of the precision measurement of $X^1\Sigma(n_v, J(=N), F, m_F) = (0,0,1/2, \pm 1/2) \rightarrow (1,0,1/2, \pm 1/2)$ transition frequencies of QH^+ (Q: even isotope of group II atoms) ions. With these molecular ions, the electron spin, electron orbital angular momentums, and Q nuclear spin are zero and the energy structure is simple. Although H nuclear spin is 1/2, there is no hyperfine splitting in the $J = 0$ state. Table 5.14. lists the transition frequencies, natural spectrum linewidth (given by spontaneous emission rate), coefficient of fractional quadratic DC Stark shift, and coefficient of fractional linear Zeeman shift, which were estimated by *ab initio* calculation [47]. The Zeeman shift is strictly linear with this transition frequency, and it is eliminated by averaging the $m_F = \pm 1/2 \rightarrow \pm 1/2$ transition frequencies (not required while aiming the measurement uncertainty of 10^{-16}). The Stark shift induced by the trap electric field (< 0.03 V cm^{-1}) is below 10^{-18}. These transition frequencies are free from the electric quadrupole shift. The fractional quadratic Doppler shift is below 10^{-17} with the kinetic energy of 1 mK.

The measurement should be performed using a cryogenic chamber, because blackbody radiation at room temperature induces the $(n_v, J) = (0,0) \rightarrow (1,1)$ or $(0,0) \rightarrow (0,1)$ transition with a rate higher than 0.02 s^{-1}, which interrupts the

Table 5.14. The transition frequencies, natural spectrum linewidth, coefficient of fractional quadratic DC Stark shift, and coefficient of fractional linear Zeeman shift of the $X^1\Sigma(n_v, J, F, m_F) = (0,0,1/2, \pm 1/2) \rightarrow (1,0,1/2, \pm 1/2)$ transition frequencies of QH^+ (Q: even isotope of group II atoms) molecular ions. These values were obtained by *ab initio* calculation [47].

	Transition frequency (THz)	Natural linewidth (Hz)	Fractional DC Stark shift (/(V/cm)^2)	Fractional linear Zeeman shift (/G)
$^{24}MgH^+$	49.02	5.5	2.4×10^{-16}	$\pm 1.2 \times 10^{-17}$
$^{40}CaH^+$	43.20	2.5	-3.1×10^{-15}	$\pm 1.3 \times 10^{-18}$
$^{88}SrH^+$	40.26	4.0	-4.8×10^{-15}	$\pm 8.7 \times 10^{-19}$
$^{64}ZnH^+$	53.18	7.8	3.9×10^{-16}	$\pm 1.7 \times 10^{-17}$
$^{114}CdH^+$	49.24	4.1	4.7×10^{-16}	$\pm 1.6 \times 10^{-17}$
$^{174}YbH^+$	42.28	1.5	-2.5×10^{-15}	$\pm 4.7 \times 10^{-18}$
$^{202}HgH^+$	58.16	6.3	2.2×10^{-16}	$\pm 2.2 \times 10^{-17}$

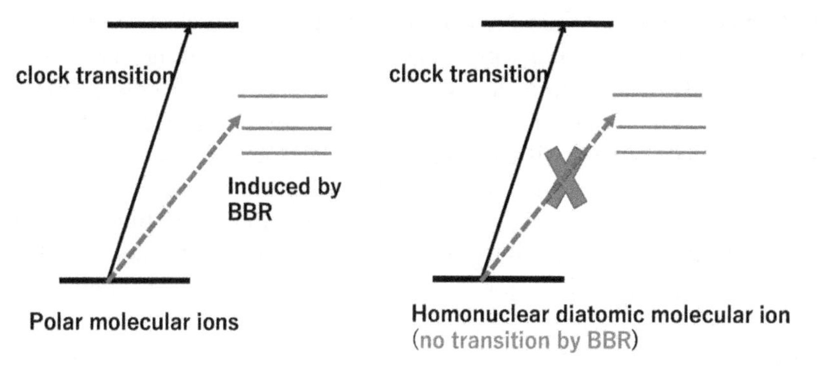

Figure 5.8. The interruption of the measurement cycle by the unhopeful transitions induced by blackbody radiation. This interruption is not caused for homonuclear diatomic molecular ions.

measurement procedure (see figure 5.8). Using the cryogenic chamber with the temperature of 4 K, the Stark shift induced by blackbody radiation is below 10^{-18}. Using a cryogenic chamber, we can prepare the molecular ion in the $(n_v, J) = (0,0)$ state with a population higher than 85% also without optical pumping.

This transition is one-photon forbidden, and a two-photon transition is required (two-photon absorption or Raman transition). Probe lasers with a high intensity (with $^{40}CaH^+$ order of 5 W cm^{-2} [48]) are required to get the two-photon transition rate higher than the spontaneous emission rate. The Stark shift induced by the probe lasers is of the order of 10^{-14}, which is eliminated using the hyper-Ramsey method [22]. The state detection is done by the quantum logical method (make an entangled state with co-trapped atomic ion as shown in section 3.3), as already performed to observe the $^{27}Al^+$ transition. The measurement uncertainty seems to be dominated by the statistical uncertainty, which is of the order of 10^{-16} measuring with a single molecular ion and the averaging period of one day. We considered only the

$n_v = 0 \rightarrow 1$ transition because the lifetimes in the higher vibrational states are shorter and the state detection by the quantum logical method is more difficult.

The $X^1\Sigma(n_v, J, F, m_F) = (0,0,1/2, \pm 1/2) \rightarrow (1,1,1/2, \pm 1/2)$ transition frequencies are also free from the electric quadrupole shift and can be a candidate for precision measurement. This transition is E1 allowed, and the experimental procedure is simpler than the two-photon transition. The probe laser with the resonant frequency pumps the molecular ion to the $(n_v, J, F) = (0,2,3/2)$ state. The state detection with the quantum logical method seems to be more convenient to monitor between $J = 0$ or 2 in the $n_v = 0$ state, because the lifetimes in both states are much longer than that in the vibrational excited state. There is a quadratic Zeeman shift in these transition frequencies because of the coupling between $(J, F) = (1,1/2)$ and $(1,3/2)$ states, and the magnetic field should be maintained to below 1 mG to attain the measurement uncertainty below 10^{-16}.

5.6.4 The vibrational transition frequencies of homonuclear diatomic molecular ions

For the homonuclear diatomic molecular ions, there is no electric dipole coupling between different vibrational rotational states in the same electronic state. Therefore, the rate of transitions induced by blackbody radiation is negligibly small at room temperature, as shown in figure 5.8. The use of a cryogenic chamber is not required for the experiment with a homonuclear diatomic molecular ion. The Stark energy shift is induced only with the coupling with electronic excited states and the Stark shifts induced by the trap electric field or blackbody radiation are four orders smaller than those for the transition frequencies of polar molecules. The lifetime in the vibrational excited states is longer than 100 s, therefore, state detection by the quantum logical method (make an entangled state with co-trapped atomic ion as shown in section 3.3) is applicable with high vibrational states. For the precision measurement, we can choose the vibrational transition $n_v = 0 \rightarrow n_v'$, which is convenient to prepare probe lasers. The natural linewidth is narrower than 10^{-6} Hz and the spectrum linewidth is determined by the linewidth of the probe laser. The probe laser with a linewidth narrower than 0.1 Hz has been attained in the wavelength range 1.4–1.6 μm using a cold Si cavity [49].

The N_2^+ $I = 0$ $X^2\Sigma(n_v, N, J, m_J) = (0,0,1/2, \pm 1/2) \rightarrow (n_v', 0,1/2, \pm 1/2)$ $(n_v' = 1,2,3,4\ldots,)$ transition frequencies are one of the best candidates for precision measurement [50]. The Zeeman energy shift is induced only by the electron spin, which has no dependence on the vibrational state. Therefore, these transition frequencies are free not only from the electric quadrupole shift but also from the Zeeman shift. The fractional Stark shifts in these transition frequencies induced by blackbody radiation with 300 K is below 10^{-17}.

These transitions are electric one-photon forbidden, but two-photon transition is possible. Najafian *et al* [51] indicated the possibility to induce the one-photon M1 transition, although the transition rate is very low. Probe lasers with high intensity are required with both methods and the Stark shift induced by probe lasers can be significant. The Stark shift induced by the probe lasers can be suppressed by the following two methods.

(1) The $n_v = 0 \rightarrow 1, 2$ transition frequencies are convenient to measure with Raman transitions using two lasers with a frequency difference of the transition frequency. When one laser induces positive shift and the other induces negative shift, the total Stark shift can be eliminated with a proper intensity ratio. For example, the $n_v = 0 \rightarrow 1$ transition frequency of $^{14}N_2^+$ molecular ion can be measured without the Stark shift using two laser lights of 541.6 THz and 476.4 THz with the same intensities.

(2) Two-photon absorption is induced using probe lasers with ultra-narrow spectrum linewidth [49]. The transition is induced with low rate (low probe laser intensity) using a long interaction time. When the linewidth is 0.1 Hz, the transition is caused with the interaction time of 1.6 s. The $n_v = 0 \rightarrow 6, 7$ transition frequencies are convenient to observe the two-photon absorption of laser light with the wavelength of 1.4–1.6 μm with ultra-narrow linewidth [49]. The Stark shift induced by the probe laser (order of 10^{-15}) is suppressed to below 10^{-18} using the hyper-Ramsey method [22, 23].

The preparation of a N_2^+ molecular ion in a selected vibrational rotational state is possible by state selective photoionization [45]. For a $^{14}N_2^+$ molecular ion, the total nuclear spin I can be 0 or 2 in the even rotational states. With $I = 2$, the energy structure with the hyperfine structure is complicated and it is difficult to prepare the molecular ion in a selected state. With the hyperfine structure, there is also a problem with the significant quadratic Zeeman shift. High resolution photoionization (resolving hyperfine structure) is required to select the $^{14}N_2$ molecule in the $I = 0$ state. The simplest method is the measurement of the $^{15}N_2^+$ transition frequency. The ^{15}N nuclear spin is 1/2 and the total nuclear spin of $^{15}N_2^+$ molecular ion is always 0 for even rotational states and 1 for odd rotational states. Table 5.15 lists the $n_v = 0 \rightarrow n_v'$ transition frequencies of $^{14}N_2^+$ and $^{15}N_2^+$ molecular ions [52].

The energy structure of the H_2^+ molecular ion is the same as that of the $^{15}N_2^+$ molecular ion and the $X^2\Sigma(n_v, N, J, m_J) = (0,0,1/2, \pm1/2) \rightarrow (n_v', 0,1/2, \pm1/2)$ transition frequencies are free from the Zeeman and electric quadrupole shifts. The fractional Stark shift induced by blackbody radiation is below 10^{-17}. However, there are technical difficulties to satisfy equation (1.5.4) with the small mass. The sympathetic cooling effect is low because the mass of the H_2^+ molecular ion is 2/9

Table 5.15. The $X^2\Sigma(n_v, N, J, m_J) = (0,0,1/2, \pm1/2) \rightarrow (n_v', 0,1/2, \pm1/2)$ transition frequencies of $^{14}N_2^+$ and $^{15}N_2^+$ molecular ions [52].

Vibrational transition	$^{14}N_2^+$ (THz)	$^{15}N_2^+$ (THz)
$n_v = 0 \rightarrow 1$	65.20	63. 02
$n_v = 0 \rightarrow 2$	129. 42	125. 13
$n_v = 0 \rightarrow 3$	192. 63	186. 32
$n_v = 0 \rightarrow 4$	254. 91	246. 60
$n_v = 0 \rightarrow 5$	316. 16	305. 96
$n_v = 0 \rightarrow 6$	376.41	364. 41
$n_v = 0 \rightarrow 7$	435.66	421.94

times smaller than that of $^{9}Be^{+}$ ion (the lightest ion that can be laser cooled). The fractional quadratic Doppler shift is 1.6×10^{-16} with the kinetic energy is 1 mK. Therefore, H_2^+ vibrational transition frequency is less advantageous for precision measurement than $^{15}N_2^+$ vibrational transition frequency. However, H_2^+ energy structure is much simpler than HD^+ molecular ion and the transition frequencies can be rigorously calculated. Therefore, the measurement of the H_2^+ transition frequencies are useful for the confirmation of the fundamentals of quantum mechanics. The $(n_v, N) = (0,2) \to (1,2)$ transition frequency was estimated to be 65.41 THz [53].

The $^{16}O_2^+$ vibrational transitions frequencies are also advantageous for precision measurement, because ^{16}O nuclear spin is zero and there is no hyperfine structure [54–56]. The electronic ground state is $X^2\Pi_\Omega$, which indicates the electron spin is 1/2 and the component of electron orbital angular momentum parallel to the molecular axis is 1. Ω is the sum of the components of electron spin and the electron orbital angular momentum parallel to the molecular axis. $\Omega = 3/2$ and 1/2 indicate that the components both angular momentum vectors in the direction of molecular axis are parallel and anti-parallel, respectively. The molecular rotation state cannot be defined with this electronic state and the total angular momentum of electron spin, electron orbital angular momentum, and molecular rotation is only defined as $J(=\Omega + \text{integer})$, as shown in figure 5.9.

The $^{16}O_2^+ X^2\Pi_{1/2}(n_v, J, m_J) = (0,1/2, \pm1/2) \to (n_v', 1/2, \pm1/2)$ transition frequencies are advantageous for precision measurement as shown below [54–56]. The transition frequencies are listed in table 5.16 [55].

These transition frequencies are free from the electric quadrupole shift. The Zeeman shift is strictly linear with the fractional coefficient of $\mp 2.5 \times 10^{-15}$ /G, which is eliminated by averaging the $m_J = \pm1/2 \to \pm1/2$ transition frequencies. The fractional Stark shift induced by blackbody radiation is below 10^{-17}. These transitions are one-photon forbidden and transition is observed by two-photon transition using probe lasers with high intensity. The Stark shift induced by the laser light in the frequency region below 10^{15} Hz is always negative. Therefore, there is no combination of two laser frequencies to induce Raman transition canceling the Stark

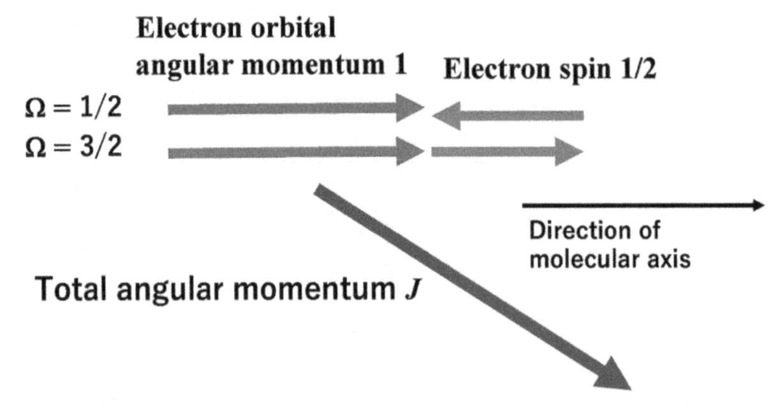

Figure 5.9. The angular moments of $^{16}O_2^+$ molecular ion in the $X^2\Pi_\Omega$ state: electron spin, electron orbital angular momentum, and total angular momentum including the molecular rotation.

Table 5.16. The $X^2\Pi_{1/2}(n_v, J, m_J) = (0,1/2, \pm1/2) \rightarrow (n_v', 1/2, \pm1/2)$ transition frequencies of $^{16}O_2^+$ molecular ions [55].

Vibrational transition	$^{16}O_2^+$ transition frequency (THz)
$n_v = 0 \rightarrow 1$	56.5
$n_v = 0 \rightarrow 2$	111.3
$n_v = 0 \rightarrow 3$	165.8
$n_v = 0 \rightarrow 4$	219.0
$n_v = 0 \rightarrow 5$	271.2
$n_v = 0 \rightarrow 6$	322.5
$n_v = 0 \rightarrow 7$	372.6
$n_v = 0 \rightarrow 8$	421.9

shifts induced by both lasers (proposed for N_2^+ transition). However, the natural linewidth of the $^{16}O_2^+$ transition is below 10^{-6} Hz and the spectrum linewidth is determined by the linewidth of the probe lasers. The $n_v = 0 \rightarrow 7$, 8 transitions can be observed by two-photon absorption of laser lights (wavelength of 1.6 and 1.4 μm), whose linewidths are reduced to below 0.1 Hz using a cold Si cavity [49]. The remaining Stark shift (order of 10^{-15}) is suppressed to below 10^{-18} by the hyper-Ramsey method [22].

The $^{15}N_2^+$ and $^{16}O_2^+$ vibrational transition frequencies can be measured with an uncertainty of 10^{-18}. The $^{15}N_2^+$ $n_v = 0 \rightarrow 7$ and the $^{16}O_2^+$ $n_v = 0 \rightarrow 8$ transition frequency is convenient to compare with the $^1S_0 \rightarrow {}^3P_0$ transition frequency of ^{87}Sr atoms in an optical lattice (429 THz), which has been measured with the uncertainty of 10^{-18} [4, 5].

5.7 Precision measurement of frequency in the THz region

The development of THz-wave source was late in comparison with microwave or optical laser sources. The THz region corresponds to the area of the vibrational rotational transition frequencies of many molecules. The establishment of a frequency standard in the THz region helps to determine the chemical structure of unknown materials including the molecules observed on other planets. The first precision measurement of transition frequency in the THz region was the $^2D_{5/2} \rightarrow {}^2D_{3/2}$ transition of a trapped and laser cooled single $^{40}Ca^+$ ion [57]. The $^{40}Ca^+$ ion is prepared in the $^2D_{5/2}$ state by the $^2S_{1/2} \rightarrow {}^2D_{5/2}$ transition induced by a laser light with a wavelength of 729 nm. The $^2D_{5/2} \rightarrow {}^2D_{3/2}$ Raman transition was produced using an optical frequency comb. The Raman transition is caused when the repetition frequency (frequency gap between neighboring frequency components) is an integer division of the transition frequency. The state was monitored using the lasers which are resonant to the $^2D_{3/2} \rightarrow {}^2P_{1/2}$ (866 nm) and the fluorescence of 397 nm is observed by the $^2P_{1/2} \rightarrow {}^2S_{1/2}$ spontaneous emission transitions. When the $^2D_{5/2} \rightarrow {}^2D_{3/2}$ transition was induced (not induced), the fluorescence was observed (not observed). The measured frequency was 1 819 599 021 534 ± 9 Hz (fractional uncertainty of

5×10^{-12}). With this experiment, the measurement uncertainty was dominated by that of the Rb atomic clock, which was used as the reference. The quadratic Zeeman shift is estimated to be 21.94± 0.02 Hz with the magnetic field of 6.5 G. The electric quadrupole shift was estimated to be −0.79± 0.02 Hz. The difference in the polarizabilities in the $^2D_{5/2}$ and $^2D_{3/2}$ states is less than 1%, and the Stark shift induced by blackbody radiation is 2 ± 6 mHz. In future, the measurement uncertainty is expected to be reduced by, for example, using ^{87}Sr lattice clock [4, 5] as the reference. This measurement method is applicable also for the $^{88}Sr^+$ or $^{174}Yb^+$ ions.

The $X^1\Sigma n_v = 0(J, F, m_F) = (0,1/2, \pm 1/2) \rightarrow (1,1/2, \pm 1/2)$ rotational transition frequencies of QH^+ (Q: even isotopes of group II atoms) and RH^+ (R: rare gas atoms) are expected to be measured with an uncertainty of 10^{-15} [58, 59]. The nuclear spins of Q and R atoms are zero. The H nuclear spin of 1/2 and the $(J, F) = (0,1/2) \rightarrow (1,1/2)$ transition frequency is free from the electric quadrupole shift. Table 5.17 lists the transition frequencies, fractional coefficients of linear and quadrupole Zeeman shifts, and the Stark shift induced by blackbody radiation. The blackbody radiation was considered with the temperature of 10 K, because the use of cryogenic chamber is required to suppress vibrational rotational transitions induced by blackbody radiation, which perturb the measurement cycle as shown in section 5.6.1.

The linear Zeeman shift is eliminated by averaging the $m_F = \pm 1/2 \rightarrow \pm 1/2$ transition frequencies.

The Stark shift induced by blackbody radiation is below 2×10^{-16}. The measurement uncertainty is dominated by the quadratic Zeeman shift, which is induced by the coupling between $(J, F) = (1,1/2)$ and $(1,3/2)$ states. The magnetic field should be maintained to below 1 mG to attain the uncertainty of 10^{-15}.

The measurement of the rotational transition frequencies is much less advantageous to obtain low measurement uncertainties than the pure vibrational transition frequencies with $J = 0 \rightarrow 0$ or $1/2 \rightarrow 1/2$ for the following reasons.

(1) The transition frequency is much lower than the vibrational transition frequency and the ratio of the frequency shifts to the transition frequencies (fractional uncertainty) is higher.

Table 5.17. List of the transition frequencies, fractional coefficients of linear and quadratic Zeeman shifts, Stark shift induced by blackbody radiation (BBR) with the temperature of 10 K in the $X^1\Sigma n_v = 0(J, F, m_F) = (0,1/2, \pm 1/2) \rightarrow (1,1/2, \pm 1/2)$ rotational transition frequencies [58, 59].

	Transition freq. (THz)	Lin. Zeeman (/G)	Quad. Zeeman (G^2)	BBR (10 K)
$^{40}CaH^+$	0.282	$\pm 7.2 \times 10^{-9}$	-1.8×10^{-9}	1.7×10^{-16}
$^{24}MgH^+$	0.382	$\mp 9.2 \times 10^{-9}$	-1.3×10^{-9}	1.1×10^{-16}
$^{202}HgH^+$	0.390	$\mp 1.1 \times 10^{-8}$	-1.2×10^{-9}	9.2×10^{-18}
$^{4}HeH^+$	2.010 183 6730	$\pm 2.4 \times 10^{-9}$	-1.8×10^{-10}	-1.7×10^{-18}
$^{20}NeH^+$	1.039 255 095	$\pm 4.1 \times 10^{-9}$	-3.7×10^{-10}	-2.6×10^{-17}
$^{40}ArH^+$	0.615 8584	$\pm 6.4 \times 10^{-9}$	-6.9×10^{-10}	-1.4×10^{-17}

(2) The Stark and Zeeman shifts at upper and lower states are quite different for the rotational transition, while they are mostly canceled for the pure vibrational transitions (same rotational states).
(3) There is a quadratic Zeeman shift induced by the hyperfine states in the $J \geqslant 1$ states.

However, the uncertainty of 10^{-15} is attained by the careful control of circumstance.

5.8 Search for the variation in fundamental constants by precision frequency measurement

The precision measurement of transition frequencies has an important role in solving the remaining mysteries in physics, because it is useful to discover a slight effect which cannot be observed with previous measurement uncertainties. For example, the precision measurement of transition frequencies of ions is useful for the search for the temporal or special variation of fundamental constants. Physical laws have been established with many fundamental constants, such as the speed of light c or the unit of electric charge. With different values of the fundamental constants, the appearance of the Universe would be quite different. For example, with a larger value of the fine structure constant $\alpha = e^2/2\varepsilon_0 hc (=0.007\,297)$ atoms with heavy nuclear could not exist because of the stronger repulsive force between protons. With smaller values of α, molecular bonding is not possible. We would not exist with any other value of α. The combination of fundamental constants seems to be too good to be a coincidence, which is one of the mysteries of physics. If fundamental constants have a dependence on time and position, we may understand that we are living in an epoch with suitable combinations of fundamental constants. In 1937, Dirac mentioned for the first time the possibility of time-varying fundamental constants [60]. But even if this hypothesis were right, its effect was expected to be too small to detect with the measurement accuracies at that time. Currently, some transition frequencies can be measured with uncertainties lower than 10^{-16} and the search for such variations has become an active topic of investigation. Recently there is an idea that the variation in the fundamental constants can be induced by the variation of the density of dark matter shown in appendix F [56].

When there is a variation in a fundamental constant X, there is a variation in the transition frequencies. Each transition frequency has different sensitivity to the variation in X. The variation in X (ΔX) is determined from the variation in the ratio of two transition frequencies $\nu_1(\propto X^{\lambda_{X1}})$ and $\nu_2(\propto X^{\lambda_{X2}})$

$$\frac{\Delta(\nu_1/\nu_2)}{(\nu_1/\nu_2)} = (\lambda_{X1} - \lambda_{X2})\frac{\Delta X}{X}. \tag{5.8.1}$$

A non-zero $(\Delta X/X)$ is detected when $\Delta(\nu_1/\nu_2) > \delta(\nu_1/\nu_2)$, where $\delta(\nu_1/\nu_2)$ is the measurement uncertainty in (ν_1/ν_2).

The variation in α has been estimated using transition frequencies of ions and neutral atoms. The sensitivity to the variation in α is high for the transition

Table 5.18. Measurement of the variation in fine structure constant α in laboratories.

α Sensitive transition	α Insensitive transition	$(d\alpha/dt)/\alpha$ (/year)
$^{199}Hg^{+}$ $^2S_{1/2}$–$^2D_{5/2}$ ($\lambda_\alpha = -3.2$)	$^{27}Al^{+}$ $^1S_0$–$^3P_0$ ($\lambda_\alpha = 0.008$)	$(-1.6\pm2.3) \times 10^{-17}$ [19]
$^{171}Yb^{+}$ $^2S_{1/2}$–$^2F_{7/2}$ ($\lambda_\alpha = -6.0$)	$^{171}Yb^{+}$ $^2S_{1/2}$–$^2D_{3/2}$ ($\lambda_\alpha = 1.0$)	$(-2.0\pm2.0) \times 10^{-17}$ [61]
$^{171}Yb^{+}$ $^2S_{1/2}$–$^2F_{7/2}$ ($\lambda_\alpha = -6.0$)	$^{171}Yb^{+}$ $^2S_{1/2}$–$^2D_{3/2}$ ($\lambda_\alpha = 1.0$)	$(-0.5\pm1.6) \times 10^{-17}$ [21]
$^{171}Yb^{+}$ $^2S_{1/2}$–$^2F_{7/2}$ ($\lambda_\alpha = -6.0$)	$^{87}Sr^{1}$ S_0–$^3P_0$ ($\lambda_\alpha = 0.06$)	$(1.0\pm1.1) \times 10^{-18}$ [62]

frequencies where the relativistic effects with the motion of electrons are significant. Table 5.18 lists the previous estimation of variation in α [19, 21, 61, 62].

Previously null results about the variation in α have been reported. However, there is still a possibility to search for the non-zero variation in α using transition frequencies with higher sensitivity. For example, the transition frequencies of highly charged ions have one order higher values of $|\lambda_\alpha|$ than that for neutral atoms or single charged ions [38].

If there is a variation in the fine structure constant α, there should also be a variation in the proton-to-electron mass ratio μ $(= m_p/m_e)$, because the proton mass m_p is 45 times larger than the total mass of the constituent particles (two up-quarks and one down-quark). The proton mass is dominated mainly by the binding energy, which is sensitive to the variation in the electromagnetic force between quarks ($\propto\alpha$). The variation in μ $(\Delta\mu/\mu)$ is given by $R_{\mu-\alpha}(\Delta\alpha/\alpha)$, where $R_{\mu-\alpha}$ is estimated to be the value between 28 and 40 from the grand unification theory (GUT) [63]. By comparing the variations in α and μ, we can obtain useful information for GUT.

The atomic transition frequencies in the optical region have very low sensitivity on μ ($\lambda_\mu < 10^{-4}$). From a comparison between the Cs hyperfine transition frequency ($\propto\alpha^2/\mu$) and the optical transition frequency, $(d\mu/dt)/\mu$ was found to be $(0.2\pm1.1) \times 10^{-16}$ per year [21], $(-0.5\pm1.6) \times 10^{-16}$ per year [61], and $(-8\pm36) \times 10^{-18}$ per year [62] respectively.

Molecular rotational and vibrational transition frequencies, given by the nuclear motion, have sensitivities to variations in μ. The length scale of atoms and molecules are proportional to the Bohr radius a_B, which is inversely proportional to the electron mass m_e. The frequency standard ν_e defined by the atomic transition frequency is given by the electric potential energy. Therefore, the following relation is derived.

$$\nu_e \propto e^2/a_B \propto m_e. \tag{5.8.2}$$

The $N \to N + 1$ rotational transition frequency ν_R is given by

$$\nu_R = 2(h/8\pi^2 I_{n_v})(N + 1), \tag{5.8.3}$$

where I_{n_v} is the moment of inertia, which is a product of the reduced mass ($\propto m_p$) and the square of the internuclear distance ($\propto a_B$). From the relation

$$I_{n_v} \propto m_p a_B{}^2 \propto m_p/m_e{}^2, \tag{5.8.4}$$

Table 5.19. List of transition frequencies to be linked aiming to search for the variation of fine structure constant α and the proton-to-electron mass ratio μ. The sensitivities for the variations in both parameters are also listed [64].

Transition	λ_α	λ_μ	Place
$^{133}Cs\ ^2S_{1/2}\ F = 3 \rightarrow 4$	2.83	−1	National Physical Laboratory
$^{87}Sr\ ^1S_0 \rightarrow\ ^3P_0$	0.06	0	National Physical Laboratory
$^{171}Yb^+\ ^2S_{1/2}\text{–}2F_{7/2}$	−6.0	0	National Physical Laboratory
$N_2^+\ X^2\Sigma n_v = 0 \rightarrow n_v'$	0	−0.5	Sussex University
$^{40}Ca^{19}F\ X^2\Sigma n_v = 0 \rightarrow n_v'$	0	−0.5	Imperial College
$Cf^{15+}\ ^2F_{5/2} \rightarrow\ ^2I_{9/2},$	47	0	University of Birmingham
$Cf^{17+}\ 5f_{5/2} \rightarrow 6p_{1/2}$	−45	0	University of Birmingham

the dependence of ν_R on μ is given by

$$\frac{\nu_R}{\nu_e} \propto \mu^{-1}, \tag{5.8.5}$$

although it is not strictly valid because of the centrifugal force. The $n_v \rightarrow n_v + 1$ vibrational transition frequency is approximately given by the harmonic vibrational frequency at low vibrational states. The vibrational transition gives the change in the electronic energy ($\propto 1/a_B$) with the change in interatomic distance ($\propto a_B$). Therefore,

$$\Delta\nu_e \propto \frac{1}{a_B} \propto m_p {\nu_v}^2 {a_B}^2,$$

$${\nu_v}^2 \propto \frac{1}{m_p {a_B}^3} \propto \frac{{m_e}^3}{m_p},$$

$$\frac{\nu_v}{\nu_e} \propto \mu^{-1/2}. \tag{5.8.6}$$

Equation (5.8.6) is also not strictly valid because the vibrational potential is not strictly harmonic. The measurement of the vibrational transition frequencies with uncertainly below 10^{-17} is useful for the search of the variation in μ.

In the UK, there is a project called QSNET 5 to establish a link to search the variations in α and μ by comparing the transition frequencies shown in table 5.19 [64].

5.9 Precision measurement of the mass of an ion using a Penning trap

The RF-ion trapping is useful for precision measurement of transition frequencies as shown above. On the other hand, a Penning trap is useful for the precision measurement of the mass ratio between different charged particles by comparing the frequencies of circle motions (see section 1.4).

For example, the ratio of the electron mass to the mass of a $^{12}C^{6+}$ ion was measured to be 4.571 499 259 × 10^{-5} [65]. The electron mass was determined with the fractional uncertainty of 10^{-10} by determination of the Rydberg constant [66] and the fine structure constant [67] from the precision measurement of the atomic transition frequencies and the definition of the Planck constant [68]. Then the masses of ions are determined after the measurement of the ratio to the electron mass. The masses of atoms are estimated by adding the masses of ions and electrons and giving the relativistic correction. The precision measurement of the mass ratio between electrons and atoms made it possible to establish the novel mass standard and the Avogadro constant.

The relation between particles and anti-particles is one of the hottest mysteries in modern physics. The characteristics of anti-particles are close to the charge conjugation + mirror image of particles (CP-symmetry). But a matter (particles) dominated Universe is not possible without the violation of the CP-symmetry. The violation of the CP-symmetry was discovered [69] and a theoretical explanation was also provided [70]. There is also the question of whether the characteristics of anti-particles are given by the charge conjugation + mirror image + time reversal of particles (CPT-symmetry). If the CPT-symmetry is violated, the fundamentals of modern physics are no longer valid. The equality of mass and the absolute value of electric charge are required for the CPT-symmetry. The equality of |charge /mass between proton and antiproton was confirmed by comparing the cyclotron frequencies, set by a magnetic field of 1.95 T, and achieved an accuracy of 7 × 10^{-11} [71].

References

[1] Kajita M 2018 *Measuring Time: Frequency Measurement and Related Developments in Physics* (Bristol: IOP Publishing) pp 1–3

[2] Kajita M 2019 *Measurement, Uncertainty and Lasers* (Bristol: IOP Publishing) pp 1–2

[3] https://en.wikipedia.org/wiki/Central_limit_theorem

[4] Ushijima I *et al* 2015 *Nat. Photon.* **9** 185

[5] Nicholson T L *et al* 2015 *Nat. Commun.* **6** 6896

[6] Berkland D J 1998 *Phys. Rev. Lett.* **80** 2089

[7] Warrington R B 1999 *Joint Meeting European Frequency and Time Forum and the IEEE Int. Frequency Control Symp.* p 125

[8] Phoonthong P *et al* 2014 *Appl. Phys.* B **117** 673

[9] Mulholland S *et al* 2019 *Appl. Phys.* B **125** 198

[10] Tanaka U *et al* 1996 *Phys. Rev.* A **53** 3982

[11] Miao S N *et al* 2021 *Opt. Lett.* **46** 5882

[12] Prestage J *et al* 2001 *Appl. Phys.* B **79** 195

[13] Heavner T P *et al* 2005 *Metrologia* **42** 411

[14] Adler F *et al* 2004 *Opt. Express* **12** 5872

[15] Dehmelt H 1982 *IEEE Trans. Instrum. Meas.* **31** 83

[16] BIPM Recommended values of standard frequencies (https://www.bipm.org/en/publications/mises-en-pratique/standard-frequencies)

[17] Kajita *et al* 2005 *Phys. Rev.* A **72** 043404

[18] Diddamus S A *et al* 2001 *Science* **293** 825

[19] Rosenband T *et al* 2008 *Science* 1154622
[20] Tamm C *et al* 2014 *Phys. Rev.* A **89** 023820
[21] Godun R M *et al* 2014 *Phys. Rev. Lett.* **113** 210801
[22] Yudin V I *et al* 2010 *Phys. Rev.* A **82** 011804
[23] Huntemann N *et al* 2016 *Phys. Rev. Lett.* **116** 063001
[24] Margolis H S *et al* 2004 *Science* **306** 1355
[25] Dube P *et al* 2017 *Metrologia* **54** 290
[26] Matsubara K *et al* 2008 *Appl. Phys. Express* **1** 067011
[27] Huang Y *et al* 2016 *Phys. Rev. Lett.* **116** 013001
[28] Dube P *et al* 2014 *Phys. Rev. Lett.* **112** 173002
[29] Champenois C *et al* 2010 *Phys. Rev.* A **81** 043410
[30] von Zanthier J *et al* 2000 *Opt. Lett.* **25** 1729
[31] Ohtsubo N *et al* 2020 *Opt. Lett.* **45** 5950
[32] Keller J *et al* 2019 *Phys. Rev.* A **99** 013405
[33] Chou C W *et al* 2010 *Phys. Rev. Lett.* **104** 070802
[34] Chou C W *et al* 2017 *Phys. Rev. Lett.* **118** 053002
[35] Brewer S M 2019 arXiv:902.07694
[36] Kimura N and Kajita M 2021 *J. Phys. Soc. Jpn.* **90** 064302
[37] Micke P *et al* 2020 *Nature* **578** 60
[38] Kozlov M G *et al* 2018 *Rev. Mod. Phys.* **90** 045005
[39] Stollenwerk P S 2018 *Atoms* **6** 53
[40] Calvin A T *et al* 2018 *J. Phys. Chem.* A **122** 3177
[41] Koelemeij J C *et al* 2007 *Phys. Rev. Lett.* **98** 173002
[42] Bressel U *et al* 2012 *Phys. Rev. Lett.* **108** 183003
[43] Biesheuvel J *et al* 2016 *Nat. Commun.* **7** 10385
[44] Alighanbari S *et al* 2018 *Nat. Phys.* **18** 74
[45] Germann M *et al* 2014 *Nat. Phys.* **10** 820
[46] Chou C W *et al* 2020 *Science* **367** 6485
[47] Kajita M *et al* 2011 *J. Phys. B: At. Mol. Opt. Phys.* **44** 025402
[48] Kajita M and Abe M 2012 *J. Phys. B: At. Mol. Opt. Phys.* **44** 025402
[49] Xiang C *et al* 2019 *Opt. Lett.* **44** 3825
[50] Kajita M *et al* 2014 *Phys. Rev.* A **89** 032509
[51] Najafian K *et al* 2020 *Phys. Chem. Chem. Phys.* **22** 23083
[52] Gilmore F R *et al* 1992 *J. Phys. Chem.* **21** 1005
[53] Karr J P *et al* 2007 *Can. J. Phys.* **85** 497
[54] Kajita M 2017 *Phys. Rev.* A **95** 023418
[55] Carollo R *et al* 2018 *Atoms* **7** 1
[56] Hannnecke D 2021 *Quantum Sci. Technol.* **6** 014005
[57] Solaro C *et al* 2018 *Phys. Rev. Lett.* **120** 253601
[58] Kajita M *et al* 2020 *J. Phys. B: At. Mol. Opt. Phys.* **53** 085401
[59] Kajita M and Kimura N 2020 *J. Phys. B: At. Mol. Opt. Phys.* **53** 135401
[60] Dirac P A 1937 *Nature* **139** 323
[61] Huntemann N *et al* 2014 *Phys. Rev. Lett.* **113** 210802
[62] Lange R *et al* 2021 *Phys. Rev. Lett.* **126** 011102
[63] Calmet X and Fritzsch H 2002 *Euro. Phys. J.* D **24** 639
[64] Barontini C *et al* 2021 arXiv:2110.05944v1

[65] Farnham D L *et al* 1995 *Phys. Rev. Lett.* **75** 3598
[66] Pohl R *et al* 2017 *Metrologia* **54** L1
[67] Parker R H *et al* 2018 *Science* **360** 191
[68] https://physics.nist.gov/cgi-bin/cuu/Value?h
[69] Christenson J H *et al* 1964 *Phys. Rev. Lett.* **13** 138
[70] Kobayashsi M and Maskawa T 1973 *Prog. Theor. Phys.* **49** 652
[71] Ulmer S *et al Nature* **524** 196

IOP Publishing

Chapter 6

Conclusion

Ions can be trapped for a long period of time in a small area using electric or magnetic fields. Observation of phenomena with trapped ions have contributed greatly to the development of physics.

Using this system, we were able to confirm the interpretation of the fundamentals of quantum mechanics; the eigenstates with different physical values can be coupled and converge to one eigenstate by this measurement. For example, a wave-packet of a single ion can be localized at two distant places simultaneously (called the Schrödinger's cat phenomenon). The entangled state between two ions (correlation between quantum states of both ions) was also realized. After attaining this state, we could measure the state of one ion and obtain information about the other ion.

The study of the chemical reaction between a trapped ion and the background atoms or molecules is also useful to enable discussion about chemical reactions in the Universe, which will give us important information to allow the possibility to find organisms on another planet. For a detailed study of chemical reactions, the trapped ions should be prepared in a selected quantum state. Recently the quantum state and the kinetic energy have also been manipulated for colliding atoms and molecules.

The RF ion trap is also useful for the precision measurement of transition frequencies of ions, because (1) we can take a long interaction time between ion and probe laser lights, (2) the quadratic Doppler shift is suppressed by laser cooling, and (3) the electric field is zero at the center of the RF-trap apparatus. A fractional uncertainty of 10^{-18} has been attained with several transition frequencies.

A Penning trap is useful for the precision measurement of the mass ratio between different charged particles, which contributed to establishing a new standard of mass and Avogadro constant. It made possible the confirmation of the equality of mass between particles and anti-particles.

doi:10.1088/978-0-7503-5472-1ch6

The ion trap also contributes to engineering. For example, the development of quantum computers using trapped ions can reduce the calculation time and electricity consumption.

This book is those starting out in ion trap research (mainly graduate course students) and the discussions are focused mainly on the fundamentals.

Appendix A

Transition between two or three states

In this appendix, we discuss the transition between two states a and b induced by the AC electric field of light in one direction $E = E_0\cos(2\pi\nu t)$. The Hamiltonian is given by:

$$H = H_0 + H'$$

$H_0\Psi_{a,b} = E_{a,b}\Psi_{a,b}$ $\quad \Psi_{a,b} = \varphi_{a,b}\exp\left(\frac{2\pi i}{h}E_{a,b}t\right)$ $E_{a,b}$: energy eigenvalues at a and b state

$$H' = \check{d}E_0\cos(2\pi\nu t) = \check{d}E_0\,\frac{[\exp(2\pi i\nu t) + \exp(-2\pi i\nu t)]}{2} \tag{A1.1}$$

$\check{d}$: electric dipole moment operator

The temporal change of the wavefunction $\Psi = a\Psi_a + b\Psi_b$ is given by

$$H\Psi = \frac{h}{2\pi i}\frac{\partial a}{\partial t}\Psi_a + \frac{h}{2\pi i}\frac{\partial b}{\partial t}\Psi_b + a\frac{h}{2\pi i}\frac{\partial \Psi_a}{\partial t} + b\frac{h}{2\pi i}\frac{\partial \Psi_b}{\partial t}$$

$$= aH_0\Psi_a + bH_0\Psi_b + aH'\Psi_a + bH'\Psi_b$$

$$\frac{h}{2\pi i}\frac{\partial a}{\partial t}\Psi_a + \frac{h}{2\pi i}\frac{\partial b}{\partial t}\Psi_b = aH'\Psi_a + bH'\Psi_b \tag{A1.2}$$

Here we assume

$$d_{aa} = \iiint \varphi_a^* \tilde{d}\varphi_a dV = 0, \quad d_{bb} = \iiint \varphi_b^* \tilde{d}\varphi_b dV = 0 \tag{A1.3}$$

Considering the product of $\Psi_{a,b}^*$ with both sides of equation (A1.2)

$$\frac{h}{2\pi i}\frac{\partial a}{\partial t} = bH'_{ab} \quad H'_{ab} = d_{ab}E_0\frac{[\exp(2\pi i(\nu_0 + \nu)t) + \exp(2\pi i(\nu_0 - \nu)t)]}{2}$$

doi:10.1088/978-0-7503-5472-1ch7

$$\frac{h}{2\pi i}\frac{\partial b}{\partial t} = aH'_{ba} \quad H'_{ba} = d_{ba}E_0\frac{[\exp(-2\pi i(\nu_0+\nu)t) + \exp(-2\pi i(\nu_0-\nu)t)]}{2}$$

$$d_{ab} = \iiint \varphi_a^* \tilde{d} \varphi_b dV \quad \nu_0 = \frac{E_b - E_a}{h}. \tag{A1.4}$$

Equation (A1.4) is simplified by ignoring the fast vibrational effect ($\propto \exp[\pm 2\pi i(\nu_0 + \nu)t]$), called the rotational wave approximation. Then we have

$$\frac{d^2b}{dt^2} = \frac{i}{\hbar}\left[2\pi i(\nu_0 - \nu)aH'_{ab} + H'_{ab}\frac{da}{at}\right] = 2\pi i(\nu_0 - \nu)\frac{db}{dt} - \frac{4\pi^2 \mid H'_{ab} \mid^2}{h^2}b$$

taking $\Delta_f = \nu_0 - \nu \quad \Omega_R = \frac{d_{ab}E_0}{h}$ (called the Rabi frequency)

$$\frac{d^2b}{dt^2} - 2\pi i\Delta_f\frac{db}{dt} + (2\pi\Omega_R)^2 b. \tag{A1.5}$$

As the general solution of b,

$$b = e^{i\pi\Delta_f t}\left[A\sin\left(\pi\sqrt{\Delta_f^2 + \Omega_R^2}t\right) + B\cos\left(\pi\sqrt{\Delta_f^2 + \Omega_R^2}t\right)\right]. \tag{A1.6}$$

Assuming $b = 0$ with $t = 0$, $B = 0$, and the formula of a using A is given by

$$a = \frac{2}{h\Omega_R e^{2\pi i\Delta_f t}}\frac{h}{2\pi i}\frac{\partial b}{\partial t} = \frac{1}{\pi\Omega_R i e^{\pi i\Delta_f t}}A\left[i\pi\Delta_f\sin\left(\pi\sqrt{\Delta_f^2 + \Omega_R^2}t\right) + \pi\sqrt{\Delta_f^2 + \Omega_R^2}\cos\left(\pi\sqrt{\Delta_f^2 + \Omega_R^2}t\right)\right] \tag{A1.7}$$

A is obtained from the condition of $\mid a \mid^2 = 1$ with $t = 0$, and the population in the b state is given by

$$\mid b \mid^2 = \frac{\Omega_R^2}{\Delta_f^2 + \Omega_R^2}\left(\sin\left(\pi\sqrt{\Delta_f^2 + \Omega_R^2}t\right)\right)^2 \tag{A1.8}$$

which is called the the 'Rabi oscillation'. When $\Delta_f = 0$, equations (A1.6) and (A1.7) are rewritten as

$$a = \cos(\pi\Omega_R t) \quad b = \sin(\pi\Omega_R t) \tag{A1.9}$$

and

$$\begin{aligned} &2\pi\Omega_R t = \tfrac{\pi}{2} \; a = b = \tfrac{1}{\sqrt{2}} \; \tfrac{\pi}{2}\text{-transition} \\ &2\pi\Omega_R t = \pi \; a = 0 \; b = 1 \; \pi\text{-transition} \\ &2\pi\Omega_R t = 2\pi \; a = -1 \; b = 0 \; 2\pi\text{-transition}. \end{aligned} \tag{A1.10}$$

We will now examine the case of the three states a_1, a_2, and b. The energy difference between the $a_{1,2} - b$ state is $h\nu_{01,2}$. When light with a frequency of $\nu_1(\nu_2)$ close to $\nu_{01}(\nu_{02})$ is considered, the $a_1 - b$ $(a_2 - b)$ transition is induced. What happens when both frequency components are irradiated simultaneously? Taking $\Psi = a_1\Psi_{a1} + a_2\Psi_{a2} + b\Psi_b$, equation (A1.2) can be rewritten as

$$\frac{\partial b}{\partial t} = 2\pi i\left[\Omega_{R1}\exp\left(-2\pi i\Delta_{f1}t + i\eta_1\right)a_1 + \Omega_{R2}\exp\left(-2\pi i\Delta_{f2}t + i\eta_2\right)a_2\right]$$

$$\Delta_{f1,2} = \nu_{01,2} - \nu_{1,2} \tag{A1.11}$$

$\Omega_{R1,2}$: Rabi frequency (defined in equation (A1.5)) between the $a_{1,2} - b$ states.

With random $\eta_{1,2}$, both transitions are induced independently. However, when

$$\Omega_{R1}a_1 = \Omega_{R2}a_2 \tag{A1.12}$$

$$\Delta_{f1} = \Delta_{f2} \tag{A1.13}$$

$$\eta_1 - \eta_2 = \pi \tag{A1.14}$$

are satisfied, $\frac{db}{dt} = 0$, and both transitions are suppressed. This phenomenon, called electric induced transparency (EIT), is then realized. Comparing the $a_1 \rightarrow b$ and the $a_2 \rightarrow b$ transition rates, the population ratio between the a_1 and a_2 states converges to equation (A1.12). When equation (A1.13) is also satisfied, both transitions are suppressed after equation (A1.14) is coincidentally satisfied after repeating the laser induced excitation and the spontaneous emission deexcitation with a random phase jump. After the transitions are suppressed, the EIT state is maintained.

As shown in equation (A1.12), the population ratio between a_1 and a_2 can be controlled by the intensity ratio of both transition frequency components. By reducing Ω_{R1} adiabatically, the state population can be localized to the a_1 state. This procedure is called ‘stimulated Raman adiabatic passage (STIRAP).

IOP Publishing

Ion Traps
A gentle introduction
Masatoshi Kajita

Appendix B

Stark and Zeeman energy shift

The electric potential $V(\vec{r})$ and the electric charge density $\rho(\vec{r})$ gives an energy shift of

$$\Delta E_V = \int \Phi^*(\vec{r})\rho(\vec{r})V(\vec{r})\Phi(\vec{r})d\vec{r}. \tag{A2.1}$$

Using a Taylor expansion of

$$V(\vec{r}) = V(0) + \frac{d}{d\vec{r}}V(\vec{r})\vec{r} + \frac{1}{2}\frac{d^2}{d\vec{r}^2}V(\vec{r})\vec{r}^2 = V(0) - \vec{E}(0)\vec{r} - \frac{d}{2d\vec{r}}\vec{E}(0)\vec{r}^2 \tag{A2.2}$$

E: electric field.

Equation (A2.1) is rewritten as

$$\Delta E_V = \Delta E_C + \Delta E_S$$

$\Delta E_C = V(0)\int \Phi^*(\vec{r})\rho(\vec{r})\Phi(\vec{r})d\vec{r} = V(0)q_e$ q_e: total electric charge

$\Delta E_S = -\vec{E}(0)\int \Phi^*(\vec{r})\rho(\vec{r})\vec{r}\Phi(\vec{r})d\vec{r} = -\vec{E}\cdot\vec{d}$ $\vec{d}$: electric dipole moment

$$\Delta E_Q = -\frac{d\vec{E}(0)}{2d\vec{r}}\int \Phi^*(\vec{r})\rho(\vec{r})\vec{r}^2\Phi(\vec{r})d\vec{r} = -\frac{d\vec{E}}{d\vec{r}}\cdot\vec{Q} \tag{A2.3}$$

Q : electric quadrupole moment.

ΔE_C denotes the Coulomb energy potential, which makes the motion of ions in the trap. ΔE_C is the state independent, therefore, it does not give any shift for the transition frequencies. ΔE_S and ΔE_Q are the Stark and the electric quadrupole energy shifts, respectively.

The electric dipole moment

$$\vec{d} = \int \Phi^*(\vec{r})\rho(\vec{r})\vec{r}\Phi(\vec{r})d\vec{r}$$

is zero for the wavefunctions Φ of atoms or diatomic molecules in the energy eigenstates with deterministic angular momentums, because the uncertainty principle of the angular momentum and the direction indicates the random direction of the dipole moment.

When there is an electric field, the formula of the Hamiltonian is given by

$$H = H_0 - \vec{d} \cdot \vec{E}, \tag{A2.4}$$

with which spherical symmetry of the Hamiltonian is no longer satisfied and the angular momentum cannot be deterministic. The eigenfunction Φ of H is given by the coupling of different eigenfunctions of H_0. For simplicity, we consider the coupling of two eigenfunctions $\Phi_{a,b}$ with eigenenergies of $E_{a,b}$,

$$\Phi = c_a\Phi_a + c_b\Phi_b \tag{A2.5}$$

$H\Phi = E_{\text{eigen}}\Phi$ E_{eigen}: energy eigenvalue.

Then

$$\Phi_a^* H\Phi = c_a E_a - c_b \overrightarrow{d_{ab}} \cdot \vec{E} = c_a E_{\text{eigen}}$$

$$\Phi_b^* H\Phi = -c_a \overrightarrow{d_{ba}} \cdot \vec{E} + c_b E_b = c_b E_{\text{eigen}}$$

$$\overrightarrow{d_{a,\,b}} = \int \Phi_a^*(\vec{r})\rho(\vec{r})\vec{r}\Phi_b(\vec{r})d\vec{r}. \tag{A2.6}$$

The non-zero coefficients $c_{a,b}$ are possible when

$$\begin{vmatrix} E_a - E_{\text{eigen}} & -\overrightarrow{d_{a,\,b}} \cdot \vec{E} \\ -\overrightarrow{d_{b,\,a}} \cdot \vec{E} & E_b - E_{\text{eigen}} \end{vmatrix} = 0. \tag{A2.7}$$

Then

$$E_{\text{eigen}} = \frac{(E_a + E_b) \pm \sqrt{(E_a - E_b)^2 + 4\,|\,\overrightarrow{d_{a,\,b}} \cdot \vec{E}\,|^2}}{2}. \tag{A2.8}$$

When $E_a > E_b$ and $E_a - E_b \gg |\,\overrightarrow{d_{a,\,b}} \cdot \vec{E}\,|$,

$$E_{\text{eigen}} = E_a + \frac{|\,\overrightarrow{d_{a,\,b}} \cdot \vec{E}\,|^2}{E_a - E_b}, \quad E_b - \frac{|\,\overrightarrow{d_{a,\,b}} \cdot \vec{E}\,|^2}{E_a - E_b}, \tag{A2.9}$$

and the Stark energy shift is given by

$$\Delta E_S = \pm\frac{|\,\overrightarrow{d_{a,\,b}} \cdot \vec{E}\,|^2}{E_a - E_b}. \tag{A2.10}$$

Equation (A2.10) shows that the Stark energy shift is inversly proportional to the energy gap between the coupled states. For the atoms and homonuclear diatomic molecules, the coupling is possible between different electric states (optical transition frequencies). For hetero-nuclear diatomic molecules, the coupling is possible between the neighboring rotational states (microwave transition frequencies), therefore, the Stark energy shift is much larger than that for the atoms and homonuclear diatomic molecules. Equation (A2.9) shows that the Stark shift is negative (high field seeking) for the lowest energy state.

With the AC electric field $\overrightarrow{E_0}\cos(2\pi\nu t)$, the number of photons with the energy $h\nu$ changes with the coupling between eigenenergy states a and b. The energy gap between two states should be considered including the difference of the photon energy. The Stark energy shift of the b state should be obtained taking the eigenenergy of the a state $E_a \pm h\nu$, and

$$\Delta E_S(b) = -\frac{1}{2}\frac{|\overrightarrow{d_{a,b}}\cdot\overrightarrow{E_0}|^2\,(E_a - E_b)}{[E_a - E_b]^2 - (h\nu)^2} = -\frac{h}{2}\frac{\Omega_R^2\nu_{ab}}{\nu_{ab}^2 - (h\nu)^2} \tag{A2.11}$$

Ω_R: Rabi frequency (see equation (A1.5)) ν_{ab}: *ab* transition frequency.

The Stark energy shift induced by a laser light (light shift) is induced mainly by the coupling between different electronic states. The atoms and molecules in the electronic ground state get a negative energy shift from the interaction with a laser light with the frequency lower than the electronic transition frequencies; therefore, ultra-cold atoms and molecules are trapped at the position where the light intensity is maximum. The laser light with a frequency higher than the electric transition frequency gives a repulsive force.

The electric quadrupole shift is significant for the precision measurement of transition frequencies of trapped ions, because of the large gradient of the trap electric field. Note that $|\vec{Q}| = 0$ when the distribution of the wavefunction is spherically symmetric (atoms with $L = 0$, molecules with $N = 0$). For the (F, M_F) state,

$$\Delta E_Q(F, M_F) \propto 3M_F^2 - F(F+1) \tag{A2.12}$$

is satisfied, where F is the hyperfine state and M_F: component of F in one direction (when nuclear spin is zero, $F = J$ and $M_F = M_J$). Average of $\Delta E_Q(F, M_F)$ for all M_F is zero. Equation (A2.12) shows $\Delta E_Q = 0$ when $F = 0$, 1/2 also for atoms with $L \neq 0$ or molecules with $N \neq 0$.

There is also the energy shift induced by the interaction between the magnetic dipole moment and the magnetic field B, which is called the Zeeman energy shift. The magnetic dipole moment is given by the electron orbital angular momentum, electron spin, nuclear spin, and the molecular rotational angular momentum. We can interpret this as the particles having spin (electron and nuclear) are like permanent magnets. The magnetic dipole moment induced by the electron orbital angular moment is like an electromagnet. Considering these effects, the total Zeeman energy shift is given by

$$\Delta E_Z = \mu_B B(g_L M_L + g_S M_S + g_I M_I + g_N M_N),$$

$$g_L = 1, \quad g_S = 2.002319, \quad g_I < 10^{-3}, \quad g_R < 10^{-4} \tag{A2.13}$$

where μ_B is a Bohr magneton ($\mu_B/h = 1.4$ MHz/G), g is the g-factor, and M is the component of angular momentum parallel to the magnetic field. Subscripts L, S, I, and N denote the electron orbital angular momentum, electron spin, nuclear spin, and molecular rotation, respectively. The Zeeman energy shift is generally not proportional to B because $M_{L,\,S,\,I,\,N}$ in energy eigenstates are not constant. We consider the wavefunction of each energy eigenstate with zero magnetic field as follows.

$$\begin{aligned}\Phi_i &= \sum p_i^k \langle M_L^k,\, M_S^k,\, M_I^k,\, M_R^k \rangle, \\ & M_L^k + M_S^k + M_I^k + M_R^k = M_{Fi}\ \text{(constant)}\end{aligned} \tag{A2.14}$$

The Zeeman shift is given by the eigenvalues of the Hamiltonian matrix with the elements of

$$\widetilde{H_{ii}} = \mu_B B \sum \left| p_i^k \right|^2 (g_L M_L^k + g_S M_S^k + g_I M_I^k + g_R M_R^k), \tag{A2.15}$$

$$\widetilde{H_{ij}} = \mu_B B \sum p_i^{k*} p_j^k (g_L M_L^k + g_S M_S^k + g_I M_I^k + g_R M_R^k) \quad (i \neq j). \tag{A2.16}$$

Equation (A2.16) gives the linear Zeeman shift considering equation (A2.15) with $M_{L,\,S,\,I,\,R} = \sum \left| p_i^k \right|^2 M_{L,\,S,\,I,\,R}^k$. The non-diagonal Hamiltonian matrix elements are shown in equation (A2.16), which induce a mixture of the different energy eigenstates and the non-linearity of the Zeeman shift (quadratic Zeeman shift with small magnetic field).

Appendix C

Motion mode of two cold ions in a string crystal

Assuming the harmonic trap potential in the q-direction ($q = x, y, z$), the equation of motion of two ions is given by

$$m_1 \frac{d^2 q_1}{dt^2} = -k_{q1} q_1 + \frac{e^2(q_1 - q_2)}{4\pi\varepsilon_0 \left[(x_1 - x_2)^2 + (y_1 - y_2)^2 + (z_1 - z_2)^2 \right]^{\frac{3}{2}}} \quad \text{(A3.1)}$$

$$m_2 \frac{d^2 q_2}{dt^2} = -k_{q2} q_2 - \frac{e^2(q_1 - q_2)}{4\pi\varepsilon_0 \left[(x_1 - x_2)^2 + (y_1 - y_2)^2 + (z_1 - z_2)^2 \right]^{\frac{3}{2}}} \quad \text{(A3.2)}$$

(A3.1) + (A3.2)

$$(m_1 + m_2) \frac{d^2 Q}{dt^2} = -k_{q1} q_1 - k_{q2} q_2 = -(k_{q1} + k_{q2}) Q + \frac{m_2 k_{q1} - m_1 k_{q2}}{m_1 + m_2} q$$

(A3.1) $\times \frac{m_2}{m_1 + m_2}$ − (A3.2) $\times \frac{m_1}{m_1 + m_2}$

$$\mu_r \frac{d^2 q}{dt^2} = -\frac{m_2 k_{q1} - m_1 k_{q2}}{m_1 + m_2} Q - \frac{m_2^2 k_{q1} + m_1^2 k_{q2}}{(m_1 + m_2)^2} q + \frac{e^2 q}{4\pi\varepsilon_0 \left[(x_1 - x_2)^2 + (y_1 - y_2)^2 + (z_1 - z_2)^2 \right]^{\frac{3}{2}}}$$

$$Q = \frac{m_1 q_1 + m_2 q_2}{m_1 + m_2} \quad q = q_1 - q_2 \quad \mu_r = \frac{m_1 m_2}{m_1 + m_2}. \quad \text{(A3.3)}$$

Here, Q and q are the center of mass and relative position in the x-, y-, and z-directions. Considering the motion of two ions in three directions, the motion modes are given by six frequency components. Here we consider the vibrational motion modes of ions in a string crystal assuming $k_{z1,2} \ll k_{x1,2}, k_{y1,2}$ ($|x_1 - x_2|, |y_1 - y_2| \ll |z_1 - z_2|$). With the ultra-low kinetic energy, the motion mode in the x,y-direction is given by the frequencies of

doi:10.1088/978-0-7503-5472-1ch9

$$\nu_m^{x1,2} = \frac{1}{2\pi}\sqrt{\frac{k_{x1,2}}{m_{1,2}}} \quad \nu_m^{y1,2} = \frac{1}{2\pi}\sqrt{\frac{k_{y1,2}}{m_{1,2}}} \tag{A3.4}$$

because x, $y \approx 0$. The vibrational motion mode of the center of mass motion in z-direction is given by the frequency of

$$\nu_m^{C-z} = \frac{1}{2\pi}\sqrt{\frac{k_{z1} + k_{z2}}{m_1 + m_2}}, \tag{A3.5}$$

and the vibrational mode of relative motion in z-direction with the ultra-low kinetic energy is given by

$$\nu_m^{R-z} = \frac{1}{2\pi}\sqrt{\frac{k_{z-\text{rel}}}{\mu_r}}$$

$k_{z-\text{rel}} = \frac{3e^2}{4\pi\varepsilon_0 z_0^3}$ where z_0 satisfies

$$\frac{m_2^2 k_{q1} + m_1^2 k_{q2}}{(m_1 + m_2)^2} = \frac{e^2}{4\pi\varepsilon_0 z_0^3}. \tag{A3.6}$$

The energy of each vibrational motion mode is given by

$$E_{m-\text{vib}}^{\alpha} = \sum\left(n_{\text{vib}}^{\alpha} + \frac{1}{2}\right)h\nu_m^{\alpha}. \tag{A3.7}$$

α: motion modes (A3.7).

The energy states of motion modes are determined by the quantum number of vibrational motion $\nu_{\text{vib}}^{\alpha}$.

This vibrational motion mode has an important role in making an entangled state of ions.

IOP Publishing

Ion Traps
A gentle introduction
Masatoshi Kajita

Appendix D

Bose–Einstein condensation

With ultra-low kinetic energy and high density, atomic waves are broadened more than the interatomic distance, and there is interference between overlapping atomic waves. Thus, the quantum state cannot be an independent event. When there are two identical particles 1 and 2, with the same quantum states at the a and b positions, the $[\varphi_1(a),\ \varphi_2(b)]$ and $[\varphi_1(b),\ \varphi_2(a)]$ states are not distinguishable, there is an interference between the two states. For particles with spin S_p, the interference of both states is given by

$$\frac{\varphi_1(a)\ \varphi_2(b) + (-1)^{2S_p}\varphi_1(b),\ \ \varphi_2(a)}{\sqrt{2}}. \tag{A4.1}$$

Particles with integer spin are called Bosons, wherein the amplitude is larger by a factor of two (probability is two times larger) when $a = b$ (the position is equal to the distance that is smaller than the position uncertainty). Bosons tend to have the same quantum states and have a condensation in the lowest energy state, which is called the Bose–Einstein condensation (BEC) state. The phase of the wavefunctions of particles is uniform and it is like a single particle with macroscopic size. We can interpret that the laser light is the BEC state of a photon.

For particles with half-integer spin, called Fermions, the wave function for $a = b$ is zero, and only one particle can exist in a single quantum state (Pauli exclusion principle). Energy distribution cannot be localized in the lowest state also with ultra-low temperature, but they can occupy the states in order from the lowest.

However, two Fermion particles with opposite spins can pair up and have Boson like states (Cooper pair). Paired Fermion atoms can also undergo the BEC state. BECs are related to the superfluidity of the helium atoms and the superconductivity (superfluidity of electrons). The motion of all particles in the BEC state is macroscopic and any scatterings of a small fractions of particles are also forbidden if there are impurities.

doi:10.1088/978-0-7503-5472-1ch10

IOP Publishing

Ion Traps
A gentle introduction
Masatoshi Kajita

Appendix E

Feshbach resonance

Feshbach resonance is the resonance between two states Φ_{atoms} (energy E_a) and Φ_{mol} (energy E_m): Φ_{atoms} is the state of atoms A and B in the ground, and Φ_{mol} is the state that atom A in an excited state is bonded with atom B. Applying the magnetic field, E_a and E_m are Zeeman shifted and both states are mixed. This mixture is significant particularly when $E_a = E_m$ (bonding energy = energy gap between two states in atom A). There is a repulsion of the crossing of both energy levels. Changing the magnetic field adiabatically, the transform between the atomic and molecular states is possible. With this method, all atoms can be transformed to the molecular state. The produced molecules are in the excited state, which can be transformed to the ground state using the induced Raman transition.

doi:10.1088/978-0-7503-5472-1ch11

IOP Publishing

Ion Traps
A gentle introduction
Masatoshi Kajita

Appendix F

Dark matter

In astronomy, there is a mystery in the form of dark matter. Dark matter is a substance that interacts gravitationally but not electromagnetically. For this reason, we cannot observe it directly.

The existence of dark matter is determined from the velocity of H atoms at the edges of galaxies. The measured velocity is much higher than expected and the gravitational force required to balance the centrifugal force is 10 times larger than that given by the matter within each galaxy. There must be some material (called dark matter) which does not have any electromagnetic interaction. The nature of dark matter remains a mystery.

The density distribution of dark matter has been measured by the gravitational lens effect (bends in the path of light from distant stars by the gravitational force). In the 1980s, the distribution of groups of galaxies was mapped and a bubble-like structure of the Universe appeared, which cannot be explained without the mass of dark matter much larger than that of matter.

doi:10.1088/978-0-7503-5472-1ch12

Lightning Source UK Ltd.
Milton Keynes UK
UKHW051454181222
414084UK00003B/52

9 780750 35470